IONS AND CAUSES

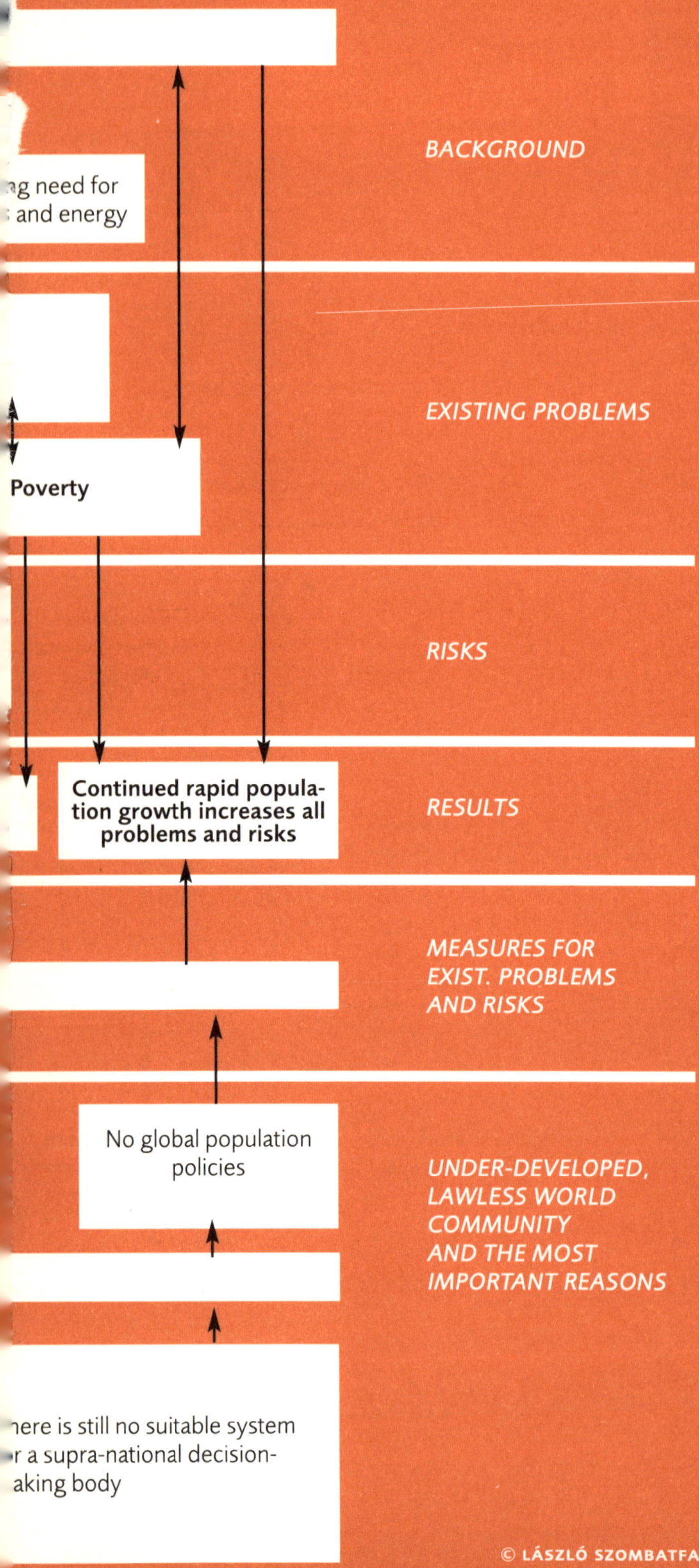

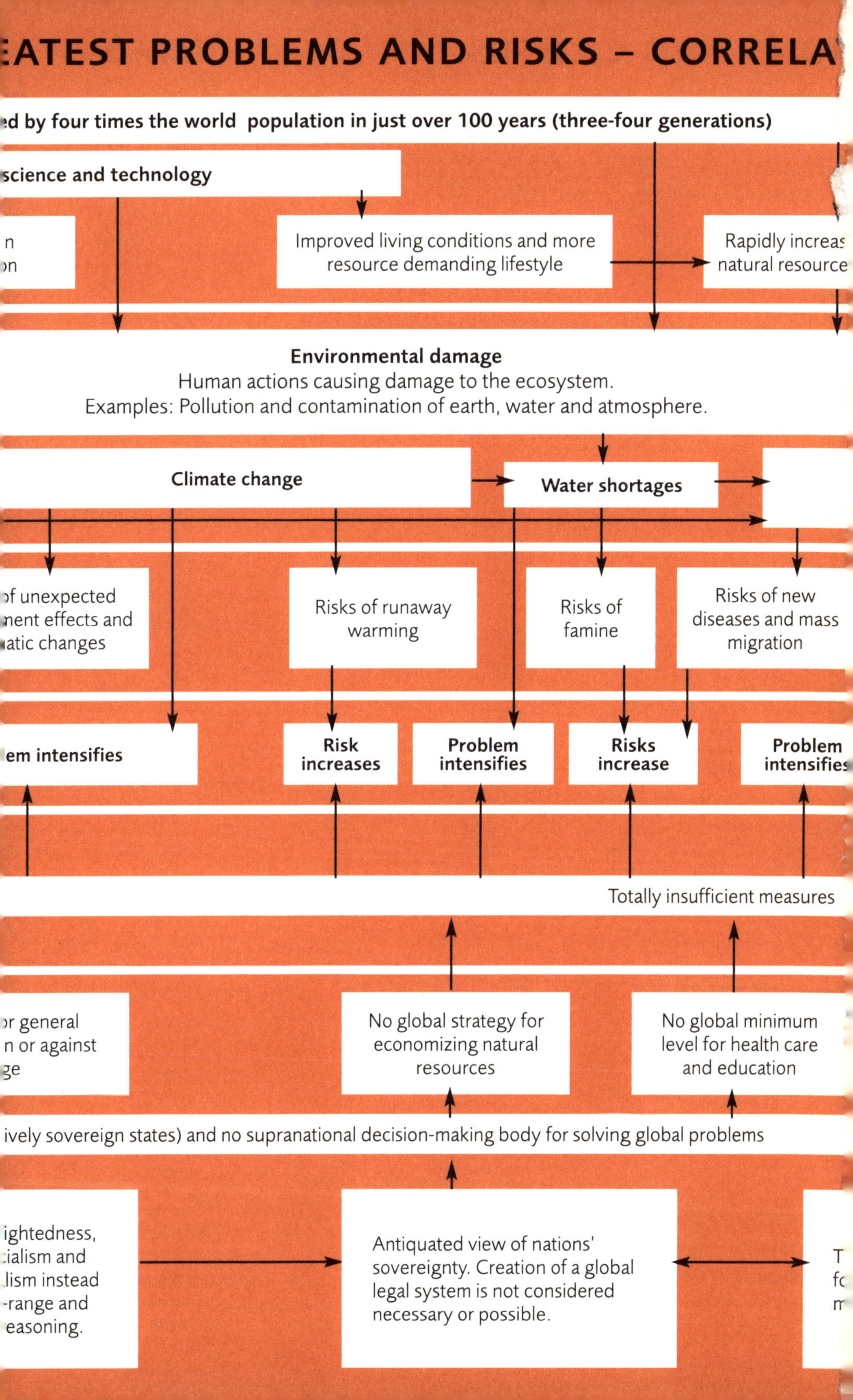

EATEST PROBLEMS AND RISKS – CORRELA
ed by four times the world population in just over 100 years (three-four generations)
science and technology
n
on
Improved living conditions and more resource demanding lifestyle
Rapidly increas natural resource
Environmental damage
Human actions causing damage to the ecosystem.
Examples: Pollution and contamination of earth, water and atmosphere.
Climate change
Water shortages
of unexpected nent effects and atic changes
Risks of runaway warming
Risks of famine
Risks of new diseases and mass migration
em intensifies
Risk increases
Problem intensifies
Risks increase
Problem intensifies
Totally insufficient measures
or general n or against ge
No global strategy for economizing natural resources
No global minimum level for health care and education
ively sovereign states) and no supranational decision-making body for solving global problems
ightedness, cialism and lism instead -range and easoning.
Antiquated view of nations' sovereignty. Creation of a global legal system is not considered necessary or possible.
T
fo
m

THE GREATEST CHALLENGES OF OUR TIME

LÁSZLÓ SZOMBATFALVY

The Greatest Challenges of Our Time

Translated from Swedish to English by Robert Skole

EKERLIDS

This book can also be read online at the following web page:
http://thegreatestchallengesofourtime.com

Ekerlids Publishing House
Tegnérgatan 37
S-111 61 Stockholm, Sweden
Ph +46 8-411 42 70, Fax +46 8-411 42 71
Mail: info@ekerlids.com
www.ekerlids.com

Graphic design: Carl Åkesson
Cover photo: IBL Bildbyrå and NASA/Visible Earth
Printed at Bulls Graphics, Halmstad, Sweden 2010
ISBN: 978-91-7092-155-1

CONTENTS

FOREWORD BY ANDERS WIJKMAN 6
AUTHOR'S FOREWORD 8

1. Unprecedented but highly risky advancements 10
2. Four mega-problems 15
3. Forgotten risk assessment 37
4. Three invisible explanatory factors 65
5. Difficult but necessary measures 76
6. Global legal system – sooner or later 90
7. Summing up 99

DIAGRAMS OCH TABLES:

Intolerable difference between rich and poor nations 33
The new world community's greatest problems and risks – correlations and causes 74–75
Population development over 2000 years 102–103
Almost 70 % of the world's population is in 11 nations + EU 104
Brief facts on the world's 195 sovereign states 105–111

FOREWORD

About 30 years ago, my phone rang. I was working with the Red Cross and the person who phoned was László Szombatfalvy. You can imagine my surprise when he told me what he planned.

"I want to help finance a film to arouse opinion against nuclear weapons and to promote a new world order," he explained. "It would be a high quality feature film to be shown at movie theaters worldwide. Cost is no problem. The goal is more important than anything else."

Our conversation was at a time when Leonid Brezhnev was in power in the Kremlin and Ronald Reagan in the White House. Young people today would have a hard time relating to this period. The threat of nuclear war hovered over the world like a dark shadow, just as global warming threatens the globe today.

László believed that I, as head of the Red Cross, could perform some magic and get a director, such as Ingmar Bergman, to accept the film assignment. Well, I did try, but did not succeed. The idea never came to fruition and eventually Mikhael Gorbatjov and Ronald Reagan met in Geneva and the threat of nuclear war abated, at least for the time being.

Many years later, László contacted me again. And again, it was a matter of significant issues facing the world. We met and talked for hours and hours on questions of world population, poverty, environmental pollution, and the climate and global warming. Since then, we've met a number of times. We got to know each other well and we are convinced that the way the world is developing, new directions must be taken. The kind of growth that dominates political thinking worldwide will destroy the ecosystem step by step and lead to a more unstable climate. The result will be nothing less than the collapse of the economic system.

This book that László has now written is about the future of mankind. He writes with deep conviction and concern. Considering his background, I hope that this book reaches many readers in business and industry as well as in politics. It is most unusual that a person who has worked in the financial sector becomes so strongly engaged in the problems of society's dark side.

László's book contains many thoughtful analyses and recommendations. His ideas on the world's legal order – or lack of legal order – are right on target. It is obvious that the system of how the world is run must be adjusted to today's reality and not, as it is now, based in how the world was after World War II.

But the most important question in the book is that of climate, as could be expected. László correctly points out the absence of risk assessment in the public debate. He makes comparisons with other fields – such as air traffic – and maintains that when it comes to climate we conscientiously avoid discussing risks, which are many times greater than those we fear in other fields. This, of course, violates the trust of mankind today, but even worse, of future generations.

László's point is that if people understood the risks they would be many times more willing to take action– and would thereby put greater pressure on political decision makers. This book is a strong appeal both for greater seriousness and honesty in the climate debate.

I sincerely hope that László's book reaches a broad audience and helps to sharpen the focus in discussions that are vital for the common future of mankind. The questions that László raises should be those we discuss most. Despite this, it is difficult to find these questions in the public debate. László's efforts are worthy of full respect. He is an excellent exponent for Harald Ofstad's maxim: "One should take seriously that which is serious."

Anders Wijkman

FORMER MEMBER OF THE EUROPEAN PARLIAMENT
FORMER ASSISTANT SECRETARY GENERAL OF THE UNITED NATIONS

AUTHOR'S FOREWORD

In the late summer of 2007, the world economy was struck with a financial crisis that quickly brought on extensive and serious consequences. The following year, several large, well-known global commercial banks crashed. To prevent a collapse of the international payments systems, governments, even those of the most capitalistic nations, were forced to nationalize some banks or provide them with tax-financed survival loans totaling historically inconceivably large amounts. Parallel with this, and in a highly unusual coordinated action, the world's central bank chiefs began a drastic shock reduction of key interest rates. This led to interest rates in many nations at the start of 2009 declining to slightly above zero, a record low. The financial crisis immediately affected the real economy, resulting in one region after another sliding into recession, causing sharply rising unemployment. Many claimed that the world economy faced its worst challenge since the depression of the 1930s. Others saw this economic crisis as the end of an epoch distinguished mainly by over emphasis on creating shareholder value and financial sleight of hand. The most pessimistic envisioned the beginning of the end of the consumption society.

As these lines are written, the economic crisis is still a grim reality for the world community. Yet, this is relatively mild in its scope, and is much simpler to solve, compared with some of the other problems facing the world.

Since leaving the stock market a number of years ago, I have devoted my thinking to questions far from the world of national economy and corporate analyses. During the most recent years, I have primarily concentrated on identifying and analyzing the greatest current problems facing mankind.

The deeper my insights have become on the subject, the more concerned I've become. The world community faces great challenges, far greater than ever before. The risks in the situation are underestimated because of poor or non-existent risk analysis. The most important factors behind the risks are not on the political agenda nor are they taken up in public discussions. In addition, there is no political global organization with understanding, power and authority to tackle these problems. When it concerns climate change, humankind is already in a risk zone.

Therefore, I have put my thoughts in print.* In order to encourage as many as possible to share my message, I have limited the length of the text. Those who still feel they don't have time to read the entire book, but who nevertheless are interested in the subject, can get a good understanding of the contents by studying the diagram on pages 74–75 carefully. This illustrates the world community's greatest problems and risks, their causes and significant effects.

I am not a scientist and thus have no deep knowledge in respective specific sub-fields. However I hope that my risk-oriented approach and my way of structuring general knowledge will increase understanding of the immediate need for radical, very long-range measures in order that the risks of truly large catastrophes will not continue to increase in a completely indefensible way.

Stockholm, June 2010
László Szombatfalvy

* *I wish to extend deepest appreciation to my friend* ***Björn Franzon*** *who not only reviewed the language in the Swedish original text but also assisted with the structure and contributed many hours in making the content more readable.*

1. Unprecedented but highly risky advancements

We must either follow our conscience, or it will pursue us.

GEORG WULF

During the past 250 years, from about 1750, developments in human history have been unparalleled. Since the start of industrialization, world population has increased 8.5 times, from 800 million to 6.8 billion. It is presently increasing at a rate of 1.2 percent annually, which might seem modest, but means there will be 12–13 billion people in 50 years!

Science and technology have advanced at record speed during this extremely short period of human history. Industrialization has basically changed society and business and industry. One can speak of a revolution in health care and in communication and the spread of information. Unfortunately, the same applies to weapons technology. Today, man can destroy human life on a scale that has never before been imagined.

During this extremely eventful period in human history, nations' economies have developed in a highly asymmetrical way. The average gross national product per person in the industrialized nations, in which one-sixth of mankind's most prosperous people live, is *80 times greater* than the equivalent figure for those in the poorest developing nations, which have roughly the same population.

Technical development has also meant that standards of living have been greatly changed, demanding far more resources and producing more waste, primarily in the richest nations. This, combined with the increase in population, has resulted in excess

Recent decades of economic growth in many poor nations has meant that large numbers of people have improved living conditions, education, health care and access to modern technology. (Photo: Thomas Wester/Scanpix)

utilization of many natural resources, as well as greater pollution and damage to the environment. Nature has been the loser during this period.

Human life and living conditions in the world's more than 190 sovereign states have become all the more intertwined with each other. National decisions and behavior now very often influence the inhabitants of other nations or even all humanity. The global community has successively begun to resemble a world enterprise – although it is still an underdeveloped establishment that lacks rules and institutions of the kind that provide a national society its values; for example, a legal system, social security, universal education and environmental protection.

An important consequence of this development is that differences between various interests have become more common and deeper, even between nations that are geographically far distant from each other. Another consequence is that it is far more important that our political leaders make correct decisions, since many more nations and many more people are affected by their decisions than in the past.

The past 250 years of humankind may have been incomparably successful in many respects, but many serious, extremely difficult problems have also been created. The greatest, in my opinion, are the following:

Environmental damage is potentially the greatest threat, since in the worse case it can make the world more or less uninhabitable for mankind.

Organized political violence strikes millions of people every year. This can also start a threat that leads to mega-catastrophes earlier than the worst environmental disasters. An increasing number of nations have nuclear weapons that can quickly wipe out large cities and even small nations. The number of people who can be killed by tomorrow's biological and chemical weapons is unknown today.

Poverty is not a direct threat to mankind, but is a phenomenon that now kills many more people and causes far more suffering than the other mega problems combined. Every third person on earth today lives in unacceptable poverty. This not only demonstrates the moral bankruptcy of rich nations but is also a destabilizing factor in the world community, as well as a risk factor for new diseases.

These three mega-problems – environmental damage, weapons of mass destruction, and poverty – and their accompanying risks have several things in common. For example, they affect each other mutually and negatively, and as a result are difficult to correct independently of each other. Another common denominator is that they cannot be solved on a national level but must be tackled through cooperation between many, or all, nations. Added to this is that none of these problems have any satisfactory solutions in sight. Finally, they can harm both present and future human generations' fundamental living conditions.

"I have no right to life's abundance
until everyone has received life's necessities."

RALPH S. CUSHAM

The earth's desert areas widen as a result of climate change. Seen here is part of the desert in Namibia. (Photo: Claes Grundsten/Scanpix)

2. Four mega-problems

> ***"If you believe that you need not do anything to solve the world's problems, you are one of them."***
>
> ANONYMOUS

Hardly anyone today would deny that the previously mentioned problems are serious and are difficult to solve. However, views differ on how serious the problems are and the risks connected to them. We shall discuss this a little deeper in this chapter.

Environmental damage can be defined, at least from our viewpoint, as human activities that affect the earth's ecological system in such a manner that in the long term humankind is more harmed than helped.

The global ecosystem – that is, nature as we usually speak of it – consists of a countless number of physical, chemical, and biological sub-systems of greatly varying sizes and complexity. All parts of the system are connected and function as an immensely complicated "machine" that is powered by the sun's radiation and some contribution by the earth's heat. Human beings are perhaps the most developed sub-component in this "machinery".

Human beings – now a threat to the ecosystem

The ecosystem is recognized as being extremely stable despite the fact that many of its sub-systems constantly change and evolve. Nature can adjust itself to slowly changing conditions but can be seriously damaged by rapid changes. Prior to industrialization, rapid changes in the ecosystem always were the results of powers outside of the biosphere: for example, volcanic eruptions or impacts of meteors or other objects from space. However, the recent unprecedented rapid changes were caused by other factors, inner forces, specifically from us humans. We have become so numerous and so technically advanced that today we humans are the greatest threat to the ecosystem.

The human capacity to utilize, affect, injure and destroy nature has unfortunately increased much faster than our knowledge of how ecological systems function. For example, we haven't even adequately discovered how we ourselves function. Just think of the many unsolved problems in medicine. Nor should we forget that there is a huge difference between gaining knowledge about humans and gaining knowledge of unique systems, such as the climate system.

Since mankind consists of a great many relatively similar varieties, one can test and study individuals without endangering the entire species. And this has been done physically and psychologically for thousands of years. But similar research cannot be carried out with unique systems. Certainly, most ecosystems do have a self-healing capability, but when it comes to unique systems one cannot continue to try to test the limits of irreparable damage without risking catastrophic disruption. Although we should know better, when it comes to environmental damage we humans are like children who play with fire without understanding the risks.

We throw sand in nature's machinery in many different ways despite the fact that we are entirely dependent on the machinery's operating effectively. One can speak of direct and indirect environmental damage. Direct damage is the result of over-utilization of renewable natural resources or conscious encroachment which damages nature. Indirect environmental damage results from unintentional – often unforeseen – side-effects of human activity.

Although climate change is a typical example of such side-effects, we will here take up this problem separately from other environmental damage. The main reason is that climate change during the past decade has been a global political issue of the highest level. In the rest of this book, we therefore no longer speak of three, but of four mega-problems facing humanity.

ENVIRONMENTAL DAMAGE – AN ENORMOUS, GROWING PROBLEM

Environmental damage (excluding climate change) stands out as an enormous and growing problem. Overfishing as well as excess use of ground water are examples of human over-utilization of the earth's renewable natural resources. Devastation of rain forests and drainage of wetlands are two examples of conscious damage to nature. Among the results are shrinking natural resources (such as depleted fish stocks); shortages of clean drinking water, which cause disease and sickness; increased release of carbon dioxide; and a lesser number of species, which reduces the stability of ecosystems.

The side effects of our activities can be foreseen and unforeseen. Consequences of our polluting or poisoning water, land and air include such effects as reduced quality of water resources, reduced or total elimination of certain species of animals and vegetation, bleached or destroyed coral reefs, unhealthy particles in air that we breathe, ozone holes in the atmosphere, and much more.

This is just a very short description of today's problems caused by environmental damage. We will not delve deeper into the specific problems, but will briefly describe examples of a major problem and a serious future threat.

Fresh water shortages

The major problem involves fresh water. It is well-known that fresh water is essential to mankind, not only for drinking but also for agriculture, hygiene and health. Globally, there is no shortage of fresh water. But it is so unevenly distributed geographically that that there is a serious shortage for some 2.5 billion people. Shortages of clean water are estimated to directly cause the deaths of about 6,000 people, mainly children, daily. In China and India,

Deforestation

Short-sighted madness

The world's forests make up a powerful and central part of the global ecosystem. In prehistoric times, forests covered about 60 million square kilometers, or 40 percent, of the earth's land surface. Up to the middle of the 1800s, the total forest area was reduced by about 10 million square kilometers. And in the past 150 years, another 10 million square kilometers disappeared, at an alarmingly accelerated rate.

The global reduction is the net effect of a slow increase of forest area in industrialized nations and a faster cutback of forests in developing nations. The worst damaged are the tropical rain forests whose area was estimated at 10 million square kilometers at the start of the 21st century, but since then is estimated to have declined by 1 percent annually because of over-harvesting – more adequately described as deforestation. The earth's rain forests are now mainly in South America (55 percent) and in Africa (23 percent).

Deforestation is obviously carried out for economic gains. Forests are cleared for agriculture – for raising soy or palm oil, etc., or to get more grazing land for beef cattle, or to provide timber to sawmills, or for roads or construction. It is a well-known fact today that these economic calculations do not take environmental consequences into consideration. All decision-makers now seem to recognize this short-sighted madness, but the negative development is allowed to continue.

The basic reasons behind the deforestation of the rain forests are local poverty and the population explosion, combined with higher living standards in many nations, which increase demand and prices for natural products.

But the environmental effects of deforestation of the rain forests are extremely alarming. Some examples:

• **Deforestation** is estimated to account for about 20 percent of the world's carbon dioxide emissions. Among other things, the harvested trees release carbon dioxide through decomposition or when burning biomass. Furthermore, the destroyed forest areas can no longer absorb carbon dioxide from the air – the rain forests' most important capability of being a so-called carbon dioxide reducer.

• **Water's life cycle** is weakened. For example, there is less rain, which is vital for agriculture and drinking water. In addition, there is the risk that the remaining forests disappear because of reduced rain combined with the forests' special sensitivity to drought.

• **Biodiversity** is depleted, weakening the entire ecosystem. It is estimated that about 50 percent of all living land species are found in rain forests. About 40 percent of the species are estimated to have become extinct since the beginning of the 1970s.

• **Negative sociological effects** are another consequence of deforestation. Native inhabitants, about 50 million living within the rain forests and some 350 million nearby, are dependent on the forests and/or clean drinking water. Deforestation creates risk that poverty worsens.

agriculture is threatened by a rapid decline of ground water levels because of irrigation methods and high demand for water for certain crops, such as rice. Climate change and the continued population growth are expected to worsen an already difficult water situation in many parts of the globe. Desalination of sea water as well as transportation of water over long distances are today too costly and thus do not appear to be a solution to the problem – at least not with existing technology.

Threatened shortages of food supplies

The threat concerns food supply – which is closely linked to the fresh water question. Food supply is not a problem today, viewed globally. The unequal distribution is caused by economical and political actions. But in the not too distant future food supplies can be threatened even in a global perspective. Fish stocks are already declining and the earth's arable land is presently utilized almost to capacity. Climate change threatens agriculture through drought and floods and other problems. Of course, the amount of arable land can be increased somewhat and more efficient agriculture and improved crop breeding can certainly continue to increase yield per hectare. But if one considers all factors it is highly uncertain whether we will be able to properly feed the earth's population if it increases from the present 6.8 billion to 9–10 billion in the coming 50 years (according to the UN's official prognosis).

Thus, under the heading of foreseeable future risks, we can add environmental damage's negative effects on our other mega-problems, among them being greater poverty and resultant conflict over natural resources.

The unforeseeable environmental consequences of human activities cannot be identified. But we must not forget that many of the known, serious, indirect effects of environmental damage were also unforeseeable – everything from bleached coral reefs and ozone holes to the more serious indirect consequences, namely climate change.

But before we move on to the climate problem, let's first

devote a few lines to the ozone hole. As is well-known, there's an ozone layer in the earth's upper atmosphere which, together with oxygen in the air, absorbs the largest part of the sun's highly-dangerous ultra-violet rays. Without this protection hardly any plant or animal life could exist, at least not on land. Two man-made compounds from the elements, chlorine and bromine, are easily vaporized, rise to the ozone layer and begin to break it down. Industrial release of chlorine has caused not only thinning of the ozone layer but holes in it. Luckily, several researchers could quickly find the reasons for the damages. And because of the imminent dangers – and because measures to correct it are relatively inexpensive – most countries have enacted laws prohibiting chlorine release. Few people, however, know how chance played a role in the drama. Paul Crutzen, a Dutch chemist who shared a Nobel Prize in Chemistry in 1995 for his research in connection with the ozone problem, said in an interview with the Swedish magazine *Forskning & Framsteg* (Research and Progress), issue Number 3, 2009: "It was pure dumb luck that industry, from the beginning, chose to use chlorine and not bromine in its products. If they used bromine, the ozone layer would have disappeared over the entire globe, not only over the Antarctic."

This statement is a valuable reminder of how important it is that we humans must investigate to the fullest possible extent the consequences of any planned activities that may affect the environment and that we must be most observant of changes in the ecosystem.

CLIMATE CHANGE WITH UNACCEPTABLE CONSEQUENCES

According to a very large majority of the world's climate experts, greenhouse gases caused by human activities have quite likely started – or at least have been the main reason for – a general continuing increase of the earth's average temperature. Emissions are largely traceable to our most important energy source, fossil fuels. This general temperature increase causes a climate change

The greenhouse effect

Global warming – a mixed blessing

The reason for the continuing global warming – which, in turn, causes climate change – is usually called the greenhouse effect. The term is somewhat inappropriate since the real greenhouse effect arises when air in an enclosed space is warmed by visible short wavelengths of the sun passing through a glass wall, or similar barrier, while the longer wavelengths from the heated objects are prevented from escaping. This is what happens in a greenhouse – or in an automobile parked for a day in strong sunshine, with light passing through the windshield or windows.

The earth's surface, of course, is not an enclosed space. The sun's warming effect on the earth corresponds to similar warming radiation from the earth to space. If this balance is upset in either direction, the result is a lower or higher average temperature of our planet. If all solar light and all warming radiation pass unhindered through the atmosphere, the average temperature would be about 33 degrees C below today's levels. That would make it minus 19 degrees C. No human, and perhaps no life, would exist.

The prevailing, much more comfortable, average temperature of around 14 degrees C is the result of the natural "greenhouse effect". This means that water vapor and certain gases in the atmosphere absorb the largest part of the heat radiating from the earth's surface and then immediately "re-radiate" it, but in all possible directions. A large part of the heat radiation is re-radiated from the so-called greenhouse gases to earth. The most important greenhouse gases are water vapor, carbon dioxide, methane and nitrogen oxide.

This natural and life-supporting "greenhouse effect" has been strengthened in the past 150 years, because human activities increased the amount of greenhouse gases in the atmosphere. Re-radiation of heat to the earth has increased while net radiation from the earth has been reduced. In other words, the balance between incoming and outgoing heat radiation has been upset and the average temperature of the earth has begun to rise. In the latest decades climate researchers have become more worried about this development, the results and risks of which are described in this book's main text.

The greenhouse effect is currently intensified by carbon dioxide (contributing 77 percent), methane (14 percent), and nitrogen oxide (8 percent).

According to the 2007 report of the Intergovernmental Panel on Climate Change, IPCC, two-thirds (or 66.3 percent) of greenhouse gas emissions are from production or use of energy, of which industry accounts for 19.4 percent, goods and passenger transport 13.1, construction 7.9, energy generation (primarily electricity) 25.9. The remaining 33.7 percent of emissions are from deforestation (17.4 percent), agriculture and ranching (13.5 percent) and waste treatment (2.8 percent).

with effects which are only partly predictable, but the probable consequences are already most alarming. *(In this book, all temperatures are in Centigrade, C.)*

The earth's average temperature in modern times, until about 1920, was 13.7 degrees C (plus/minus 0.3 degrees C). After that, the average temperature has clearly been rising and until 2005 climbed by 0.75 degrees C. Carbon dioxide content in the atmosphere has increased faster, from about 280 ppm (parts per million) during the 1700s to close to 390 ppm today. Looking at the total greenhouse gas concentrations – of which there is no data for the 1700s – the level is about 450 ppm. Even without any additional increase, this level will gradually lead to a total increase of average temperature by 2 degrees C.

This may seem like a moderate warming, but it has already noticeably changed the earth's climate and created risks of serious damage to the ecosystem. And if we humans fail to take action, but continue as before, global warming – according to experts – will likely increase by more than 5 degrees C. in about 100 years. And to make matters worse, the foreseeable damages are expected to increase proportionally more than the temperature. If we do not change our emissions practices, risks of catastrophic consequences for all mankind will increase faster.

The question is why such relatively modest changes of average temperatures, between 14 degrees C and up to 20 degrees C, are expected to have such catastrophic consequences, when humans are accustomed to live in and enjoy normal lives in temperatures ranging between minus 5 degrees C and plus 40 degrees C?

Major consequences of small changes

One explanation is that warming will not be the same globally, but just the opposite. An increase of the average temperature by 0.75 degrees C can cause an increase of 2–3 degrees C in some areas, as has occurred in Siberia. Another explanation is that a modest general warming can lead to serious climate changes. It is easier to understand how only a few degrees higher temperature can essentially change our living conditions when you consider

Climate change

Not a bolt out of the blue

As far back as 100 years ago, Swedish scientists observed that human activities could affect the climate. Arvid Högbom, professor of geology in Stockholm, warned in 1895 that anthracite burning would increase carbon dioxide content in the air. The following year, Svante Arrhenius, professor of physics and Nobel Prize Laureate, estimated that doubling of the content of carbon dioxide in the atmosphere would lead to an increase of the earth's average temperature by 5-6 degrees C. However, with the low emissions at that time, the process would take several thousand years.

In 1938, measurements by Guy S. Callendar, an English researcher, confirmed theories that the amount of carbon dioxide in the atmosphere had actually increased since the previous century. His report made little impact since attention at that time was focused on the outbreak of World War II.

During the 1950s and 1960s, several research reports were published supporting Svante Arrhenius's calculation of carbon dioxide emissions' warming effects. But the time perspective in these reports has been reduced considerably.

In the 1970s, it was discovered that emissions of several other greenhouse gases from human activities heightened carbon dioxide's effects.

In 1988, the International Panel on Climate Control, IPCC, was organized. Every fourth or fifth year since 1990, the IPCC has published climate change reports that are increasingly more extensive and ominous.

In December 1997, the first international agreement to limit emissions of greenhouse gases was signed in Japan. Known as the Kyoto Protocol, the agreement's goal is that industrialized nations reduce emissions of greenhouse gases by 5.2 percent by 2012, compared with 1990 levels. The Protocol has been ratified by 176 nations, but unfortunately not by the USA or China.

that the earth's average temperature during the last ice age was only 5 degrees C below today's level.

Four distinct consequences of global warming

What are the consequences of a general increase in temperatures? Unfortunately, the answer to this question contains many uncertainties today. But based on existent knowledge we can identify four distinct consequences of a general increase in temperature.

- Various climate effects generally linked to rising temperatures.
- Serious consequences of large scale melting of ice.
- Major, irreversible changes in climate and the ecosystem, (the so-called threshold effects, or tipping points).
- Runaway warming – an extreme threshold effect – meaning that the temperature increase itself releases mechanisms in nature that results in self-sustaining, continuous rising temperatures, entirely out of human control.

Let's discuss each of these consequences, one by one.

Various climate effects

Various climate effects generally linked to increased temperatures have been widely discussed. These, for example, include extreme storms and cloudbursts, drought, expanding deserts, heat waves and tropical species spreading out, both north and south.

Partly obvious, partly probable, the consequences of these effects are also well-known: for example, declining or devastated agriculture; flooding; more forest fires; problems with food supplies; starvation; fresh water shortages; increased poverty; worsened health; new diseases; forced migration; more refugees; armed conflicts within and between nations; material damage; economic and political crises, both national and global. In other words: deterioration of mankind's living conditions. And the poorest nations will be hurt first.

To all this, can be added the reduction or depletion of biodiversity, which, in itself, contributes to the weakening of the ecosystem's ability to withstand disturbances. At only a 2 degree C rise in temperature, up to 40 percent of all species are threatened with extinction. (For additional information, the "must" source to consult is *The Economics of Climate Change,* The Stern Review, 2006.)

Large scale melting of ice

Ice melting and the rise of sea levels are the most obvious consequences of increased temperatures.

Melting ice and a rise of the sea level are the most obvious consequences of a rise in temperatures. (Photo: Scanpix)

Melting glaciers

Fresh water supply in danger

Glaciers are melting faster than expected. This can lead to both rising sea levels and serious shortages of fresh water. According to Tim Barnett, marine physicist at Scripps Institution of Oceanography, University of California at San Diego, up to two billion people worldwide are entirely or partly dependent on water melted from glaciers. How is this so?

According to the Swedish National Encyclopedia, a glacier is an accumulation of snow and ice that moves because of its own weight. At a thickness of 30 meters the top layer is so heavy that its pressure exceeds the firmness of the inner ice. This causes a slow deformation process – creeping – in the ice mass and a glacier is formed. The earth's glaciers were formed during the latest ice age.

When dangers of glacial melting are discussed in connection with climate change, they usually refer to the slightly less than 2 percent of the world's glaciers that are outside the Polar regions. (If the Antarctic and Greenland glaciers – which are within the Polar regions – were to melt away it would cause the world's seas to rise dramatically.) Glaciers in the temperate climate zones are only located in high mountains, where the sun, even in the summer, doesn't melt away all the snow and ice. Several of these glaciers therefore serve as gigantic water reservoirs for humans and animals. In the winter, the glaciers collect and freeze precipitation which thaws during the summer. The melted water runs down into rivers and keeps them from drying out in the dry season. People living along these rivers are periodically wholly dependent on melted glacial water for drinking and agriculture.

If glaciers were to melt away due to global warming two successive effects would result. During the melting period the abnormal amount of water could cause rivers to overflow. And when the glaciers entirely melt away there would be no supply of water to rivers in dry seasons, and this would leave people living in large areas without drinking water for short or long periods. The glaciers of the Himalayas alone periodically supply seven of Asia's largest rivers, rivers that several hundred million people in China and India depend on. Water shortages would also threaten people in other parts of the world. For example, areas of Peru are dependent on water from the Pastoruri Glacier in the Andes. If this glacier were to melt away, water supply to Lima, the capital city, would be threatened.

Melting glaciers would also have other negative consequences. For example, when the glaciers' ice cover begins to disappear, the newly exposed dark ground surface would absorb more solar radiation, which contributes to general rising temperatures.

It should first be pointed out that climate change causes sea levels to rise not only from melting ice but marginally also directly through so-called thermal expansion. Water – like most other matter – expands in volume when heated.

In discussing this subject, one usually distinguishes between three different types of ice: sea ice, glaciers and other land ice.

Melting sea ice: Melting of sea ice has no effect on sea levels but can negatively affect living conditions of polar animals. In addition, it contributes to the general warming through the so-called Albedo Effect. This means that the dark waters that are exposed when ice melts absorb much more solar radiation than white ice.

Melting glaciers of Greenland and the Antarctic increase sea levels, but only marginally. The negative consequences primarily affect humans living along the rivers that are supplied by water flowing from the melting glaciers. These people are generally dependent on water flowing from the melting glaciers during the summer dry season. During the winter, snow rebuilds the glaciers' ice that was lost in melting. But when a glacier melts down over a number of years due to climate change, the extensive melting first causes rivers to overflow. When the glacier melts away entirely, there's a water shortage in the dry summer periods, which can be catastrophic for hundreds of millions of people along the rivers.

Melting land ice in Greenland and the Antarctic is the absolute greatest threat to sea level. An increased sea level can certainly damage coastal areas above sea levels, because salt water ruins arable land and drinking water. But the greatest threat obviously is flooding that unmercifully forces evacuation of humans and causes enormous economic losses.

It is known that a rise in sea level by as little as one meter, or by just several meters would be enough to make hundreds of millions of people homeless and without land to reclaim. Two-thirds of the earth's population live in coastal areas, while half of all humans live within 200 kilometers of coasts. It is estimated that as early as 2025 a total of 75 percent of the earth's population will live in costal areas. If the major cities on the coast, home to many millions of people, were destroyed by rising sea levels, the world economy could be severely devastated in the short term.

Experts predict that if only Greenland's ice melted this would increase sea level by 7 meters, while melting of the west Antarctic ice would raise the level by 5–6 meters. How many billions of people would be forced to flee from their homes? And to where? An unimaginable refugee situation would be created and the world community would be faced with enormous rescue and resettlement efforts.

If the entire Antarctic ice were to melt away, sea levels would rise by at least 64 meters! Even if such a catastrophe would require a thousand years or several thousand years to develop, and although it is foreseeable and unavoidable, it would cause extremely serious damage to mankind.

Tipping points – major irrevocable changes

Among the future risks associated with climate change exist also other threats. Climate researchers often use the term "tipping points" to describe various threshold effects. These irrevocable events in the ecosystem are expected to happen – or are expected to begin to happen – when a specific temperature increase is reached and are expected to cause major, permanent changes in the climate system or in mankind's surroundings.

All the threshold effects cannot be specifically determined. Outside of the previously mentioned melting of Greenland and Antarctic ice, the following events are generally considered predictable tipping points:

- The vital Gulf Stream could be severely weakened or would come to a standstill.
- The Amazon rain forests could dry out and turn into savannahs – flat, tropical grasslands.

- The so-called El Niño weather periods would be more common and much stronger, which would lead to additional warming, more dry weather in the inlands of Southeast Asia and South America, greater flooding along South America's coasts, and devastating consequences for the Amazon rainforests.

These climate events would lead to such radical changes in the earth's climate that it is most difficult to speculate on the consequences based on today's knowledge. Hundreds of millions of people, perhaps more than a billion, would probably be forced to seek new places to live. It is both easy and yet impossible to forecast what such a process would mean for the global political situation and for international and national conflicts.

Nightmare scenario Number One

But there is an even more dangerous threshold effect, namely that which is generally called runaway global warming.

Humanity's nightmare scenario Number One means that nature itself takes control of continuing warming. This process, for example, could be started by ice melting and releasing sufficient amounts of the enormous quantities of methane gas that is enclosed in ice crystals in the tundra of the Polar areas and in sea sediment. Once free, this would initiate a horrific spiral of more methane emissions, greater amounts of greenhouse gases, higher temperatures, more melting, more gas emissions, and on and on. Methane in a 100-year perspective is 23 times a more powerful greenhouse gas than carbon dioxide and, in this way, would make global warming self-propelled, entirely out of human control.

Nobody knows what level of temperature increase would cause runaway global warming. This is why predictions about the consequences of such warming are so uncertain. But it is likely that the aforementioned threshold effects (and probably even more) would be triggered if the average temperature increases by more than 6 degrees C. Billions of people would be forced to flee if the East Antarctic's enormous ice mass also melted, partly or entirely. The world map would certainly have to be re-drawn and most likely not much of the globe would look the same.

> ***"Why should I care about future generations? They have never done anything for me."***
>
> GROUCHO MARX

POLITICAL VIOLENCE – ALWAYS PRESENT

The third mega-problem is that of organized political violence, which here will be called simply political violence. This includes war, civil war, genocide, forced migration, terrorism and other organized violence with a political objective.

Political violence is a never-ending problem, one that causes countless deaths and all kinds of other human suffering. Since the end of World War II we have not experienced a single year without political violence occurring in various locations around the globe. Millions of people of all ages have been killed, injured, raped or driven from their homes. And most of the victims have been civilians, women and children.

Besides the direct horrors, the side-effects have been horrendous. War and civil war in developing nations worsen poverty and can seriously injure humans psychologically – both the victims and the perpetrators. For example: how great are the chances that child soldiers grow up to enjoy a normal life?

Looking at economic damages, besides the direct physical destruction in the wake of war, we must consider the costs of human and material resources devoted to military purposes. According to a report from the Stockholm International Peace Research Institute, the total global cost in 2008 was 1,454 billion US dollars. And how much is this in comparable terms? About 2.7 billion, or the poorest 40 percent of mankind, live on almost exactly this total amount. And the annual global aid to developing nations is about 100 billion dollars, or only 7 percent of the amount spent on "defense".

Threat of weapons of mass destruction

The greatest risk of political violence is that weapons of mass destruction will be used in political conflicts or by terrorists. This risk becomes even more serious as additional nations gain the knowledge and technology to produce nuclear warheads and chemical and biological weapons.

The general public became conscious of the enormous dangers of these weapons at the end of World War II when atomic bombs were dropped on Hiroshima and Nagasaki in 1945. Today's weapons of mass destruction are far more horrendous and can destroy major cities with populations in the millions and even small nations, or spread life-threatening diseases (epidemics and pandemics) among the civilian population.

In addition, political violence and its risks consume considerable time and attention of political leaders – and worse – they draw the attention of politicians and the public away from other important problems. This can mean that measures to solve life-threatening problems are taken too late or not at all, which, in turn, increases risks of other catastrophes.

In summary: Tremendous problems connected with political violence exist right now, and the future risk is clearly many times larger.

POVERTY – THE WORST AFFLICTION TODAY

Poverty is obviously an extremely relative concept. In Sweden, the minimum level of existence is generally accepted as 150 kronor, or about 20 dollars, per day. Internationally, poverty means something quite different: that a person can live on 2 dollars a day – or 1.25 dollars (which is classed as extreme poverty). In addition, those living in poverty lack basic health care. Poverty, as discussed in this book, can be defined as the condition in which the lack of basic material resources and impoverishment harm or endanger peoples' physical or mental health.

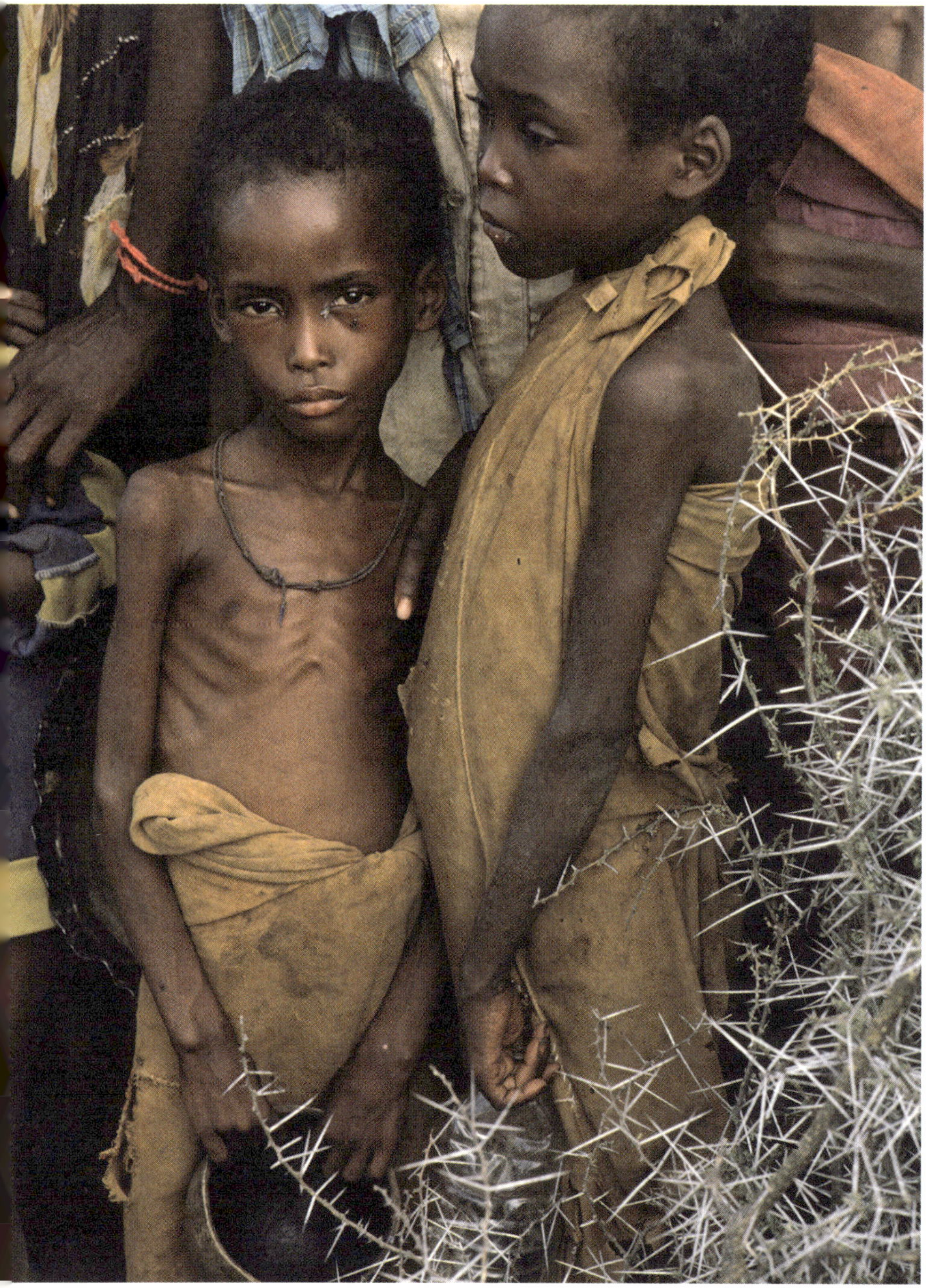

Poverty is the worst cause of suffering today. More than 2.5 billion people now live on two dollars a day, at the most. Photo taken in Somalia.
(Photo: Wesley Bocxe/IBL Bildbyrå)

Intolerable difference between rich and poor nations

This book names poverty as one of mankind's four mega-problems – just as it is the problem that causes the absolutely greatest human suffering today. The table below, based on an August 2009 database from the Swedish Institute of International Affairs, shows frightening differences between the richest and the poorest nations. The world's population in this table is divided into three groups: the richest, middle and poorest nations. The richest and poorest groups each account for 15 percent of the total world population.

The gross national product, GNP, per person in the richest nations is over 80 times higher than in the poorest nations! Note that the difference refers to the average values for the respective groups. The difference between the individual wealthiest nations and the poorest nations are naturally even more challenging.

Not only are the differences dramatic but the specific economic levels are just as striking. The poorest group, with a combined population of almost one billion, has a per capita GNP of only 472 dollars annually, or $1.30 per day. This is not even one-hundredth of Sweden's equivalent figure: $55.620 dollars, or $152 per day.

The data also illustrates economy's significance for life expectancy, educational levels, population growth and carbon dioxide emissions.

People in the rich nations live an average of 40 percent – or 23 years – longer than those in the poor nations. The connection between poverty (undernourishment, inadequate health care, and education) and rapid population growth are also clearly seen in the table.

The table's figures for the three groups' carbon dioxide emissions per person clearly show the extremely difficult problems facing world leaders.

Nations	Population share of world total (%)	GNP 2008 per person (USD)	Average life expectancy* (years)	Average literacy (%)	Annual, natural population increase (%)	CO2 emission per person and year (metric tons)
Richest	15	38 719	79	99	0.36	12.4
Medium group	70	2 920	68	81	1.14	3.0
Poorest	15	472	56	58	2.26	0.4
Maximum, Minimum values						
Rich nations		118 045	83	100	-0.7**	36.9
Poor nations		113	30	22	3.9	0.1

Source: Swedish Institute of International Affairs, database, August 2009
Data for carbon dioxide refers to 2005
** Average for women and men*
*** Information refers to Ukraine, which here is considered neither among the rich nor poor nations, but is in the Medium group*

Moral bankruptcy

According to the World Bank, about 1.4 billion people – about 20 percent of humanity – live today on a maximum of 1.25 dollars per day, while an equal number live on maximum 2 dollars per day. The average living costs for the two groups are obviously much lower. That a dollar has greater purchasing power in poor nations than in the USA has already been taken into account in these estimates. In detail, among other things, this means:

- about 900 million people starve, or, to use a more refined expression, are chronically undernourished
- about 1.1 billion people lack clean drinking water
- about 2 billion people lack hygienic toilet facilities
- because of poverty, every fifth child dies before his or her fifth year, which means that 10 million small children die annually, or 27,000 per day. More than half of these die of malnutrition and almost every fifth child dies of dehydration or diarrhea because of impure water.

This deplorable accounting could be much longer. Of the four mega-problems discussed in this book, poverty is the one that currently causes the incomparably greatest human suffering – a fact that indicates a moral bankruptcy of rich nations which fail to take sufficient measures to reduce poverty in developing nations.

For poverty-stricken mankind, the future risks will primarily mean even greater suffering due to shortages of food and water, and more political violence.

From the viewpoint of the wealthier nations, the greatest risks are linked to population growth's environmental damage, mass migrations, the spread of new and old diseases, and destabilization of political systems.

Paralyzing insecurity

In this chapter, we have attempted to briefly show that our four global problems are extremely serious and that several of the potential risks are enormous, and even catastrophic.

However, neither the public nor political leaders seem to view the situation in this way – judging from the non-existent demands from the public to take speedy and effective measures, or judging from politicians' failure to make decisions on these issues. This is especially clearly seen when it comes to greenhouse gas effects. Climate change can drastically worsen mankind's living conditions, yet greenhouse gas emissions have been allowed to increase every year – despite the Kyoto agreement of 1997 and all the negotiations about it in the following 12 years. Since the Climate Convention was signed in Rio de Janeiro in 1992 – in which the nations of the world for the first time agreed on stabilizing the climate – emissions of greenhouse gases have increased more than 30 percent.

There seems to be great uncertainty, especially among the public, about whether the threats as described are real or are highly improbable nightmare visions. This uncertainty can greatly limit politicians' possibilities to take necessary, but unpopular, measures.

The reason for this uncertainty is probably largely due to the fact that no investigations have yet been made – or at least not published – on the probabilities, the risks, that the enormous catastrophes will occur. This is very hard to understand, since risk assessment is normally the most important component in forming a basis for decisions in uncertain situations – even when far smaller values are at stake compared with vital questions of environment, climate, poverty and political violence.

The next chapter is devoted to an attempt at risk assessment based on publicly available data.

"Compassion in action may be the glorious possibility that could protect our crowded, polluted planet."

VICTORIA MORAN

A rise of only one or more meters of the sea level would be enough to make hundreds of millions of people homeless. The 1,192 islands of the low-lying Maldives (photo) are already in the risk zone. (Photo: Scanpix)

3. Forgotten risk assessment

> ***"We bear responsibility not only for what we do but for what we fail to do."***
>
> R. WHITLEY

There is considerable uncertainly about climate change consequences and risks. The same can be said about the three other challenges facing mankind – environmental damage, political violence, and poverty. Therefore, all important decisions aimed at solving these problems should first be subject to risk assessments. However, as pointed out in the previous chapter, risk assessments are conspicuous by their absence in the public debate.

Risk and its two components

Just what is meant by risk assessment?

If we start with the basic concept, we mean risk as a negative event that would injure or damage our interests and that can occur in the future with a certain – known or unknown – probability. Risks therefore consist of two components: the potential (possible) damage itself and the probability that the damage will occur.

In everyday usage, risk is often meant to describe only one of the two components, sometimes the first, sometimes the second, but also a combination of both, or the total risk. For example, if we say that someone would take a great risk if he wants to jump down to a stone floor from five meters height, what we mean is either that he can severely hurt himself, or that there is a high probability that he will hurt himself, or both of these. In this book, when we discuss risk without quantifying the probability in percent or proportions, we will always mean the total risk. We will always mean the potential damage when we say damage. There are risks of various types, among them those that we consciously take because we know that the chances involved can be more than well compensated for. A familiar example is investing in stock market mutual funds instead of depositing money in secure bank savings accounts. Other risks can either unfairly hurt us or be brought on by ourselves, unconsciously or because we do something in spite of the risk. Risks involved in the ongoing climate change are of the latter type. We would naturally eliminate or at least reduce such risks – as long as it is well worth the cost.

Eliminating or reducing risk is seldom free. As a rule, we must give up something in return. This can either mean that we sacrifice something we have (as when we pay for home fire insurance instead of using the money for something else) or when we refrain from something we would like to have or like to accomplish (such as a heavy smoker deciding to stop smoking to reduce the risk of lung cancer).

In order to make these decisions we must measure, weigh or evaluate partly the risk (or the reduction of the risk we are considering) and partly the self-sacrifice we must make to reduce or eliminate the risk. This procedure – carefully trying to analyze the risk in order to proceed in a rational way – is what I label risk assessment.

Variable importance of probability

The amount of risk thus depends partly on the possible amount of damage and partly on the probability of damage occurring. When it involves ordinary economic damage, such as a house fire, risk is equal to the potential damage multiplied by the probability of a fire. Since risk is always less than the possible damage, a 50 percent probability is expressed as 0.5. If a newly built single-family home would cost 1 million dollars to rebuild after total destruction by fire and the estimated annual probability of such total damage is one in 1,000, the risk is 1,000,000 x 0.001 = 1,000 dollars. This would be equivalent to the theoretical annual insurance premium. (In reality, there would be additional charges for the insurance company's costs and calculated profits.)

But – and this is a central point – this simple risk equation does not apply when the damage cannot be repaired or replaced by money, for example, when it involves human life or permanently reduced quality of life. In such cases, the amount of risk is more dependent on the amount of possible damage. Extremely large potential damage would make probability insignificant. A mathematician would say that infinity multiplied by 0.00000000001 is very infinite.

Translated to everyday life: When we know that we can lose something that is irreplaceable we do everything in our power to avoid the loss. An example can clarify this reasoning. Assume that two parents in a rather poor nation learn that their beloved little child suffers from a rare illness which, without a specific, very costly, treatment, causes death in three out of 10 cases, while those treated recover completely. But this treatment is found only abroad, in a rich nation. The parents work and earn decent wages and live well but have no savings. They cannot imagine exposing their little girl to a deadly risk if there is any way they can avoid it. Therefore, they borrow money to pay for all costs of treatment abroad, despite the fact that this means that the family will have to live at a much lower standard than they have been accustomed to for the next five years (when the loan is paid off).

The most interesting question of principle here is this: What would the parents have done if the probability that the child would have died without treatment was not 30 percent, but 5 percent, or only 1 percent? The mathematical risk would then have been only one-sixth part or one-thirtieth part compared with the original assumptions. Would the parents want to eliminate the risk and would their willingness to make sacrifices be reduced to an equal degree? Hardly anyone would believe they would. The parents knew that there was a risk and that they could have bad luck and suffer the improbable. But if that happened, they would recognize that they would have to live knowing that their unwillingness to make sacrifices – that is, their prioritization of material well-being – caused the child's death.

Negligible and non-negligible risks

Obviously, we must accept certain risks if we are to live a normal life. If the possible damage is insignificant, we usually don't bother thinking about risks. And in some situations, we accept very small probabilities that there will be very large damages because we feel that the risk is negligible. In this context, we mean that negligible risk has such small likelihood of any eventual damage happening that people in general will act as if the probability of damage is equal to zero, despite the fact that the potential damage could be very large. This often involves reoccurring risks that we feel are negligible in view of the advantages of taking the risks. For example, most of us do not even think of risks of commercial flying or driving in a car since we want to lead a normal life. An important point in risk assessment is to distinguish between negligible and non-negligible probability.

It is quite natural that aversion to risk – like everything else – differs considerably from person to person. To get an idea of the limits of negligible risks, or the public concept of risks, we can look more closely at air and automobile travel, two important activities that directly reveal how society and individuals tolerate accident frequency.

During the period 2001–2008, there were a total of 239 airplane accidents, or about 30 per year. During the same period there were a total of 25 to 30 million flights globally. This means there was about one crash per one million flights. This very low probability (in numbers, 0.000001) perhaps has less to do with risk tolerance than reliable technology. Let us assume, however, that the probability of a tragic airplane crash is not one in one million flights, but one in 1,000 or one in 100. We would then read of about 2,500 or 25,000 crashes per month, or 8 or 80 per day. If the world statistics were representative for the USA, we would then have 7,000 or 70,000 airplane crashes out of the 7 million flights taking off from American ground annually. Would there still be any commercial passenger air traffic? Not at all likely. Neither public authorities nor potential passengers would accept such risk levels, even if they were "only" one per thousand.

It's quite possible that air traffic would have developed and

Correction

An unfortunate misprint occurred toward the bottom half of page 40 where the author compares the risks involving climate change and air travel.

The sentence citing the comparative figures should have stated: "We would then read of about 2,500 or 25,000 crashes per month, **or 80 or 800 per day.**" (And not 8 or 80 per day.)

With this crucial correction it becomes even more strikingly clear that the risks mankind appears to accept regarding climate change is anything but negligible.

Deforestation decreases the world's forests at an alarming rate. Worst hurt are tropical rain forests, resulting in serious environmental consequences. The photo shows a devastated rain forest in Costa Rica. (Photo: Peter Hoelstad/Scanpix)

existed even if it had not succeeded in attaining its present safety levels. Nobody knows what the highest accident frequency authorities and passengers would have accepted, but it's unlikely the threshold would have exceeded ten times the level accepted today. That would have meant about 300 accidents per year globally, or almost one a day or one accident every 100,000 flights.

Auto accidents in Sweden each year kill about 500 people and seriously injure some 5,000. If each of Sweden's 9 million people took a car trip only once a year, this would mean a death or serious injury for every 1,700 car trips. But since each Swede makes an estimated 50–100 car trips each year, a fair estimate is that the risk level is one accident per 100,000 car trips. Actually, the risk with the average car trip is much smaller since intoxicated drivers and speeders are highly over-represented in the statistics.

Using the previous reasoning, we can come to the un-scientifically-based but quite reasonable conclusion that the limit for negligible probability – when possible deaths are limited to between one and several hundred – is somewhere between one per 100,000 and one per 1,000,000.

Eliminate, reduce or limit

If we accept that the probability of an accident is negligible, that's the end of our worries. Otherwise, we must try to determine the size of the risk, partly on the basis of the size and type of potential damage, and partly on the basis of the estimated or presumed probability that damage will occur.

The next step is to examine what we can do about the risks. Several alternatives are usually possible. Can we eliminate, reduce or limit the potential damage (for example, use cars equipped with air bags)? Or can we eliminate, reduce or limit probability that there will be an accident (for example, by driving very carefully)? Or can we do both?

Before deciding which measures to take, we must naturally consider the cost of the measures. We must always look at three factors: Which measures will eliminate, reduce or limit risks involved in the situation? What are the expected results of the measures? And what must we sacrifice in exchange? These are specific questions that must be studied case by case.

What are reasonable sacrifices?

What are the sacrifices that we consider reasonable? In principle, there is only one general rule: It can't cost more than it is worth. This principle is easy to follow when there are quantifiable, financial risks, but unfortunately it is much more difficult to apply in other cases. The key factor in major, irrevocable potential damages is whether the probability of damage can be considered as negligible or not. When it is not negligible, all sacrifices are reasonable when they are smaller than the potential damage.

In practice, there can be considerable disagreement on this question because of large individual differences in the willingness to take risk, assessment of the possible damages, valuation of the sacrifices demanded and – not the least – various perceptions of the moral aspects.

SPECIFIC RISK ASSESSMENT – CLIMATE THREAT

Following this introduction to the ABCs of risk assessment, it's time for some practical application of the theories. We begin with the most complicated problem, climate change.

First, let us emphasize that we are grappling with extreme uncertainties. The public's knowledge of the global ecosystem (or nature) is very limited. Even climate experts' estimates (for example, on the consequences of greenhouse gas emissions) are very uncertain in many aspects. Nevertheless, these estimates form the basis of both the public's and politicians' understanding of and decisions about climate questions. But we have no better foundation for decision-making.

Added to the picture is the fact that experts' calculations have so far often underestimated the problems. Most of the forecast deviations reported in the media have pointed in the wrong direction. Sea ice and glaciers melt faster, sea levels increase more, the seas of the world absorb less carbon dioxide than expected, and so on. We can also refer to two of Sweden's most respected climate experts, Professors Erland Källén and Markku Rummukainen, who, in April 2009, provided additional climate information after the 2007 publication of the IPCC's major report. (IPCC, the Intergovernmental Panel on Climate Change, was established in 1988.) Among other things, the two professors pointed out that the effects of global warming are more powerful than what had been generally believed, that future changes in the climate can be greater than what had previously been shown, and that it can be more difficult than previously estimated to limit global warming to a maximum of 2.0 degrees C.

Potential damages

As previously noted, incalculable damage, small and large, can follow in the wake of climate change. In general, we can expect more serious damage the higher the earth's temperature rises.

Our risk assessment will focus on the foreseeable, major potential damages, those which can seriously worsen mankind's living conditions. As pointed out in the previous chapter, these damages are caused by certain irrevocable changes in the climate system, changes which in some cases can lead to runaway, self-propelling warming. Thus, we can verify that the potential damages resulting from climate change are extremely great, even if we disregard any potential, unforeseen catastrophic scenarios.

Probability of catastrophic climate changes

What is the probability that enormous climate catastrophes will occur? This is the most important question to try to answer. What the public believes about this will largely determine the measures that are "politically possible" to take.

But the question of the probability of catastrophe is also the most difficult to try to answer, since it contains so much genuine uncertainty and involves such a long time-frame.

Despite these difficulties, scientists must attempt to estimate how large the probability is that human greenhouse gas emissions will cause catastrophes. It will be a multi-stage assessment, with each stage, unfortunately, having special sources of error.

Three key questions

We should try to find answers to the following three key questions:

1. Which temperature increases can trigger catastrophic events?

The foundations for estimates are uncertain and opinions differ because, among other things, mankind has no historic experience of global warming and its various consequences.

2. What concentrations of greenhouse gases in the atmosphere will lead to these critical temperature increases?

Uncertainty is due to three factors:

- The strength of the greenhouse gases' general warming effect, which is also known as climate sensitivity.
- Certain feedback in the wake of general warming (that is, the consequences that strengthen or weaken the original change, in this case, increased temperature). This does not change the greenhouse gas amount but nevertheless influences temperature. For example, the darker sea or land areas exposed when ice melts absorb much more energy than the previous ice cover, which reflected most of the solar radiation.
- Certain industrial emissions which infuse the air with various types of very small, short-lived, and unhealthy particles – aerosols – which can both raise or lower solar radiation's warming effects.

3. What amounts of emissions lead to these greenhouse gas concentrations?

The uncertainty depends primarily on known and unknown feedbacks that increase the amount of greenhouse gas in the atmosphere. For example, thawed permafrost which releases methane and carbon dioxide. Or when the seas' normal ability to absorb carbon dioxide from air is reduced because of warmer climate. Such feedbacks have so far been considered only partly in experts' climate models.

These three uncertainties, when considered together, explain to a great degree the differences of experts' opinions in the climate question, as does the wide time range experts often estimate for the various effects.

Some concrete estimates are now available of the probabilities of when the irreversible climate occurrences (the tipping points) will affect the earth. *The Proceedings of the National Academy of Sciences of the US* in March 2009 published a survey of leading climate researchers' estimates of the probability of when five serious climate events will occur, based on various temperature increases from the present time to 2200. Despite the alarming estimates, the survey did not receive the publicity it deserved. One reason may be that the publication published the results of the survey in the form of rather complicated bar charts. The following table – in my opinion – shows the most interesting results of the survey, re-written into numbers for the sake of clarity.

The most ominous figures in the table are those showing that the scientists forecast as most probable that several, very serious changes in the climate system will occur even if the general temperature increases only up to 2.7 degrees C. As is well known, a temperature increase of not over 2 degrees C is the goal for climate measures that the world's leading politicians in recent years have attempted to bring about.

An increasing number of observers, however, are doubtful that this goal can be reached. This means that there are two circumstances, independent of each other, which indicate the probability is not negligible, but that instead it is quite evident that the specified climate events will occur. This is partly indicated by the probability of "below 2.7 degrees C" and "between 2.7 and 4.7 degrees C" warming. The estimated probability of "below 2.7 degrees C" is not much less than the probability of the 2 degrees C warming maximum aimed at by politicians. It's an open question about what's more realistic:

that politicians succeed or fail in reaching agreement on measures that truly limit global warming by 2 degrees C.

Irrespective of what one believes about the reliability of these experts' estimates or of the reasoning applied, the conclusion drawn from the table must be that probability is far from negligible that extremely serious changes in the ecosystem will occur.

Ominous figures

As the next step in risk assessment, we will attempt to find an indication of how great the probability is of the most feared scenario: that greenhouse gases in the thawed permafrost become the fuel for runaway global warming. Here, we can note that there are already reports of increased release of methane, both from the tundra and from the sea floor. Most climate researchers seem to avoid making their views public on what general temperature increases would be needed to cause methane to be released in such quantities as to cause runaway warming. However, a Russian permafrost researcher, Sergej Zimov, who has headed a research station in Siberia for almost 30 years, and who is considered a world authority in the field, said in an interview with the Swedish daily *Dagens Nyheter* in November 2007: "If the average temperature increases by more than five degrees the entire permafrost will inevitably melt away. But there is a risk that even at two degrees we will reach the critical point." That statement is all the more worrisome since warming has been well above average in the regions where frozen methane is found. In Siberia, for example, the average temperature has already increased by 3–4 degrees C, while the earth's temperature, as previously mentioned, has increased by 0.75 degrees C.

The central significance of probability in connection with risk assessment motivates another table of figures. The main part of these figures was published in the original edition of the well-respected Stern Review (*The Economics of Climate Change*, The Stern Review, 2006).

Unfortunately, the table was replaced in the summary of the

Results of a survey of 43 respected climate experts on the probability of five large, irreversible climate events before year 2200 which will cause dangerous climate changes.

FIVE CLIMATE EVENTS	**Number of experts** who consider themselves qualified to evaluate respective events	**Mean value for evaluating** probability for events occurring if total warming until 2200 is		
		below 2,7°C	**2,7–4,7°C**	**4,7–8,7°C**
Gulf Stream slows by at least 80 %	22	6 %	17 %	34 %
Greenland's ice melts totally or almost all	15	20 %	43 %	67 %
West Antarctic ice cap melts and slides into the sea	15	18 %	29 %	49 %
Amazon rain forests: at least half die or burn up	14	17 %	31 %	45 %
El Niño periods become more common and stronger	14	8 %	19 %	31 %

For each event a minority of the experts (between 7 and 29 percent) believed that either events would not happen or would not lead to dangerous climate changes. In estimating the above mean values for probability, these experts' evaluations are counted as zero percent probability.

The full report, "Imprecise probability assessment of tipping points in the climate system," is available on the Internet at www.pnas.org

Stern Review with another, less illuminating diagram. Only the summary was translated into Swedish, which means that there can't have been very many in Sweden who have studied the figures.

Much greater risk than flying

The table below shows various levels of greenhouse gas in the atmosphere, and the long-term temperature increases caused by the various greenhouse gas levels – according to leading climate researchers. The figures in columns 2 and 3 are from two independent research groups. The figures in the last column are from IPCC, the UN's climate panel. The researchers in this climate panel are highly respected, but political forces in nations like China and Saudi Arabia seem to put pressure on researchers to tone down climate threats in the published results.

The first column in the table shows the various levels of greenhouse gases in the atmosphere. (The measure used is CO2e, carbon dioxide equivalents, which means that all greenhouse gases are converted to carbon dioxide.) The Stern Review of 2006 gave the current concentration of greenhouse gases at about 430 ppm (parts per million), while a report in 2007 by the Swedish Science Council on Climate Change had the figure at about 450 ppm. The concentration is forecast as rising annually by 2–2.5 ppm. All indications point to our already having reached the 450 ppm level, or will reach it in the very near future. The bold-faced line in the table represents the starting point, which can be compared with the table's starting value, 280 ppm, from the mid-1700s, just prior to the industrial revolution. However, it should be noted that in the 1700s, only carbon dioxide content was measured.

Because of the climate system's own inertia, the increased concentration of greenhouse gases influences the temperature successively and over a long time. Columns 2, 3 and 4 show the various research groups' forecasts of results of this drawn-out process. This shows how much the temperatures will increase (compared with the pre-industrial epoch) at various stabilization levels of greenhouse gas content.

According to studies in 2004 by the Met Office Hadley Centre for Climate Change, even the present levels of greenhouse gases will successively cause temperatures to increase by 1.7–3.7 degrees C. The next column, (a synthesis of 11 studies published in 2006) forecasts that the long-term temperature increase – if the present level of greenhouse gases remains constant – will be in a wider range, by 0.8–6.4 degrees C. The figures from IPCC are lower. Here, the present level of greenhouse gases is expected to lead to a long-term temperature increase between 1.4–3.1 degrees C, with 2.1 degrees C as the most likely outcome.

It must be emphasized that the probability intervals in columns 2 and 3 make up only 90 percent of the possible outcome. The limits for the lowest and highest 5 percentage points are not considered. IPCC's figures are even vaguer, since they cover only 66–90 percent of the possible outcome. These figures are very important in a risk assessment, where all possible events must obviously be considered.

The signals we get from the figures in this table are as worri-

Connection between greenhouse gas concentration and anticipated warming

Anticipated probability interval for long-range temperature increases from pre-industrial levels at various stabilization levels of the greenhouse gas content in the atmosphere.

Greenhouse gas concentration		**Anticipated warming according to following sources:** *Hadley Centre 2004* (degrees C)	*11 Studies 2006* (degrees C)	*IPCC 2007* (degrees C)
280 ppm*	around yr. 1750	0	0	0
400 ppm		1.3–2.8	0.6–4.9	–
450 ppm	**present level**	**1.7–3.7**	**0.8–6.4**	**1.4–3.1 (2.1)**
550 ppm		2.4–5.3	1.2–9.1	1.9–4.4 (2.9)
650 ppm		2.9–6.6	1.5–11.4	2.4–5.5 (3.6)
750 ppm		3.4–7.7	1.7–13.3	2.8–6.4 (4.3)
1 000 ppm		4.4–9.9	2.2–17.1	3.7–8.3 (5.5)

* Only carbon dioxide.

some as those from the previous table. Despite the fact that the figures in the reports obviously differ from each other, they agree that there is a significant probability that today's levels of greenhouse gases in the atmosphere will lead to a long-term temperature increase by over 2 degrees C – and even much more than that.

What does all this indicate for the probability of runaway global warming? Let's assume that the Russian permafrost researcher Sergej Zimov's forecast is far too pessimistic and there is only a 50 percent risk that the disastrous process begins at a temperature increase of 5 degrees C. We further assume that the critical temperature increase lies between 2 and 8 degrees C, and within these limits the probability is divided approximately in accord with a normal distribution curve, which means that the process is most likely to start around 5 degrees C and least probable just over 2 degrees C and just under 8 degrees C. We can now roughly determine what the three different estimates in the table say about the probability that a catastrophe will occur at a consistent greenhouse gas concentration of 450 ppm CO_2e: according to the Hadley Centre's figures the risk is over 5 percent; according to the 11 studies, it is over 25 percent; and according to IPCC, around 2 percent.

The figures in this as well as in the previous tables must be interpreted using both common sense and a critical attitude. The apparent precision which the figures indicate must obviously be disregarded. Instead, the magnitude of the figures is decisive. Here, we must differentiate between negligible and non-negligible probabilities. Since neither 25, 5, nor 2 percent probability is negligible in this question, the various figures do not influence the magnitude of the risk nor, essentially, the measures motivated by them.

But obviously we should not conclude that it does not matter if the probability is 2.5 or 25 percent nor that there is no motivation to make great sacrifices to minimize or reduce a risk when possible, for example, by bringing a 20 percentage risk to a 5 percentage risk. This would be completely wrong!

Let's for a moment return to the example cited previously, of the parents who warded off the risk that their sick little girl would die when they took a very large loan to finance expensive treatment in a foreign land. The most interesting question here is whether the parents would have acted the same way if the expensive treatment only reduced the deadly risk from 30 percent to 10 percent, instead of as in the case given previously, from 30 percent to no risk at all, that is, to a complete recovery? I believe that the correct answer with highest probability would be yes, the parents would take the same measures. By borrowing the large amount, the parents reduced the probability of something – their child's death – that they wanted to avoid at any price.

No matter how we look at the uncertainties in current opinions of the climate problem, the conclusion is that probabilities are far from negligible that large, irreversible climate events will take place. Not even the probability of runaway warming – the "nightmare scenario" – is negligible.

The probability of these disasters would not even be considered as negligible if they only were one-hundredth as large as they appear today. If we were to say the risks for climate catastrophes according to experts are only one in 100, and then accept an objection saying there is only one percent probability that the experts are correct, this would make the risk of a world catastrophe only one in 10,000 (100 x 100), which is far greater risk than we accept for automobile or aircraft accidents.

This should be an eye-opener for all those who support the argument that current global warming is part of so-called natural climate variations and has little to do with human emissions of greenhouse gases. These opponents of actions to curb climate change demand proof that measures that require sacrifices will really slow down global warming. Naturally, this cannot be proven with 100 percent certainty, but the views of an overwhelming majority of climate researchers make it highly probable. And since the doubters can hardly insist that the probability is over 99 percent that the expert majority is wrong, they should admit that the risks of climate catastrophes are not negligible and therefore measures to minimize these threats are strongly justified.

Hurricane Katrina, which devastated a large part of the Mexican Gulf coast of the USA in 2005, caused tremendous damage and suffering for residents. The photo shows part of New Orleans. (Photo: Scanpix)

“ It is not only surprising but very frightening that the probability of extremely large catastrophes is at such a dramatically higher level than that of car or airplane accidents. ”

If we optimistically believe that the probability of climate catastrophes is only one in 100 and compare this with automobile accidents where the risk is about one in 100,000, we see that climate risks are still one thousand times greater.

It is not only surprising but very frightening that the probability of extremely large catastrophes is at such a dramatically higher level than that of car or airplane accidents.

How great are the total risks?

We have only discussed the probability aspects and have not considered the differences in potential damage from climate catastrophes compared with automobile accidents. When it comes to climate, all mankind's continued existence is endangered, compared with the risk that “only” one or a few people are killed or injured in an auto accident.

When we consider possible damage, the potential catastrophic results and the high probability of events indicate without any doubt that climate change is the greatest threat that mankind has ever faced. And more worrisome, risks are not as far in the future as most people believe.

Since the general public does not realize the true dimensions of the risks, it's not surprising that there is so little pressure on political leaders to solve climate problems.

Can we eliminate or reduce these risks?

The answer to this question is also discouraging. As we stated at the beginning of this chapter, risks can, in principle, be eliminated in two ways: either by preventing damage from occurring or by reducing to zero the probability that damage will occur. In the same way, risks can be reduced in two ways: either by reducing possible damage through preventive measures or by reducing the probability that damage will occur.

But when it comes to climate catastrophes, the potential damage can hardly be reduced to such a degree that it is no longer considered catastrophic. For example, if the sea level rises so much as to make New York or Hong Kong uninhabitable, economic damage can be limited somewhat through good planning. But not the amount of total damage.

The outlook is especially grim when it comes to the possibilities for reducing the probability of serious potential climate catastrophes. In both the short and long terms, the possibilities are equal to zero.

Mankind already in the risk zone

Many say that we both can and should reduce the risk of climate catastrophes by reducing emissions. This sounds fine, at first. But those who say this must consider the comparative risks after emissions are reduced with the risks that would have existed if we did not take any measures or simply continued to increase emissions. I maintain that a reduction by definition must mean phasing down emissions from existing levels and not from assumed or possible future levels.

So far – despite the Kyoto Protocol and a deeper insight into the possible and probable effects of climate change – humans have increased emissions of greenhouse gases into the atmosphere. This increases the risks. According to a 2007 report by the Swedish Scientific Council on Climate Issues, the concentration of greenhouse gases presently increases by 1–2.5 ppm annually. As is seen in the research report referred to in the table entitled "Connection between greenhouse gas concentrations and expected warming", the present greenhouse gas content in the atmosphere can eventually lead to the warming of the earth's average temperature by 3–6 degrees C. And, unfortunately, there is some probability of an even greater temperature increase, since the figures in these reports only cover a maximum of 90 percent of the possible outcome. This means that mankind is already in a risk zone and that we cannot eliminate these risks even if we could succeed in halting further emissions!

On top of all this, risks increase every day with every additional ton of greenhouse gas emission.

Naturally, it's impossible to completely stop emissions in the short and long term perspectives. Therefore, the best we can do is to slow down the rate of increase as much as we can when it comes to risks. And this can be done in only one way – by reducing emissions as much as possible.

When we achieve an emission level that no longer increases greenhouse gas content in the air, we will have reduced the risk increase to zero. Only after that, by further limiting emissions, will the lengthy period of restoration to health begin, with the reduction of both greenhouse gas content and risks.

Our risk assessment of climate change brings us to the following conclusions:

- Potential damages are exceptionally large.
- Probabilities that the potential damages will occur cannot be forecast with precision, but they are definitely not negligible.
- Total risks are therefore extremely large.
- Risks cannot be eliminated.

“This means that mankind is already in a risk zone and that we cannot eliminate these risks even if we could succeed in halting further emissions!”

- In the short term, not even an increase of these risks can be halted.
- The most we can do to minimize the risks is to reduce greenhouse gas emissions as quickly and as much as is reasonable in consideration of the "costs".

In Chapter 5, entitled "Difficult but necessary measures", we take up the last element in a traditional risk assesssment, that is, weighing the expected results of measures that can be taken against the sacrifices that the measures demand.

The remaining part of this chapter will be a very concise risk assessment of the other great challenges facing mankind.

SPECIFIC RISK ASSESSMENT – ENVIRONMENTAL DAMAGE

The potential devastation resulting from environmental damage is estimated to be enormous, especially because supplies of food and water for humans are in real jeopardy.

The probability is very high that catastrophes would result from environmental damage since the world community has no supranational organization that could immediately decide on and carry out extensive and effective measures to curb population growth and halt environmental damage.

When both the potential damage and probability are recognized as very great, then the risks of environmental damage are also clearly seen as very great.

It is presently impossible to even eliminate the most serious risks of environmental damage since both population growth and climate change strongly appear to worsen the problem. We can only – when it applies to climate change – make every effort to limit the increase of risks as much as possible without unreasonable sacrifices.

SPECIFIC RISK ASSESSMENT – POLITICAL VIOLENCE

Organized political violence is the greatest threat to mankind after climate change and environmental damage. The potential damage and harm must be considered as extremely great.

The probability that these potential damages will occur is hard to determine, but is by no means negligible. There is little sense in trying to apply figures to the probability of a third world war. But violence and threats breed fear and hate, and both these passions are very dangerous, especially when exploited by leaders of nations with nuclear weapons. This combination – and especially combined with some bad luck or with the help of the so-called human factor – can wipe out a significant part of mankind. As is well known, it is the human factor that usually causes the failure of even the most sophisticated and meticulously designed technological developments and security systems. And today's political systems, at both national and international levels, are anything but perfectly designed and constructed, while at the same time they are entirely dominated by the human factor.

The spread of weapons of mass destruction

The risk of political violence also continually increases – like that of environmental threats – because of technological developments and proliferation of weapons of mass destruction. The number of

“ ... violence and threats breed fear and hate, and both these passions are very dangerous, especially when exploited by leaders of nations with nuclear weapons. This combination – and especially together with some bad luck or with the help of the so-called human factor – can wipe out a significant part of mankind. ”

Weapons of mass destruction after World War II

Twice at the brink

Following World War II it was obvious that mankind must do everything possible to avoid a third world war. With the development of the atomic bomb, no nation could any longer feel secure from the consequences of large-scale war. Over 60 years later, we are forced to conclude that nuclear weapons have been made more powerful, more effective and become even more terrifying. This also applies to chemical and biological weapons of mass destruction, which are sometimes referred to as "the poor man's nuclear weapon."

Chemical weapons of mass destruction (mustard gas and nerve gas) have been used after World War II on a small-scale in some local wars and terrorist attacks, as in the Tokyo subway attack in 1995. As late as October 2009, there have been reports that China twice found traces of the deadly nerve gas sarin in the air at the border to North Korea, which resulted in reinforced border security.

On the positive side, neither nuclear nor biological weapons have been used during the post war period. However, the world came very close to the brink of nuclear war on two occasions.

The first of the two incidents was the Cuban Missile Crisis in October, 1962, when the USA discovered that Cuba had almost finished building launching pads for nuclear-armed medium distance missiles, aimed at America. American warships surrounded Cuba and prevented Soviet ships carrying missiles to reach the island nation. President John F. Kennedy demanded that the launching pads be dismantled and he warned that a missile fired from Cuba would be considered as a Soviet attack on the USA and would be answered with massive retaliation. The Soviet Union backed down, and the launch pads were dismantled on the US promise that Cuba would not be attacked.

The second incident was on September 26, 1983, when relations between the USA and the Soviet Union were again extremely tense. At a military surveillance center outside Moscow, an alarm suddenly sounded and the computerized security system reported that five intercontinental missiles were heading from the USA toward Soviet territory. According to instructions and routine, Lieutenant Colonel Stanislav Petrov, in charge of the radar center at the time, should have forwarded the alarm, which could have resulted in launching of retaliatory missiles. But Petrov didn't act, since he assumed that the USA wouldn't launch an attack with only five missiles. He believed the alarm was caused by a technical error, and thus didn't forward the alarm signal. He was right. It was later determined that solar rays hitting clouds could be mistaken for missiles due to faults in the Soviet security system. Petrov's good judgment saved the situation and probably prevented a large-scale nuclear war. But because he failed to follow regulations, he was pensioned ahead of time. The incident was first made known in 1998.

The threat of weapons of mass destruction is not only monumental but totally unpredictable. When Albert Einstein once was asked which weapons would be used in a third world war, he is said to have replied, "I don't know, but the fourth will be fought with clubs and stones."

A growing number of nations have nuclear weapons, which can destroy major cities and entire small nations. The photo was taken in Hiroshima, Japan, destroyed in 1945 by the first atomic bomb.

nations with nuclear weapons has successively increased – despite all international negotiations and agreements. Access to weapons of mass destruction in nations that are also controlled by religious leaders adds a special risk dimension. Such leaders can place greater emphasis on their believers' prosperity in the after-life than during life, which completely differs from the world view that prevailed in the east and west powers during the Cold War days. Risks must therefore be considered very great and, in addition, their timing is unpredictable.

Can we eliminate the risks of war in which weapons of mass destruction are used?

No, we cannot do so with our current world order. Until now, discussions have centered partly on nuclear disarmament and partly on limiting additional proliferation of nuclear weapons. There have been no results in either case, and anything else would have been a real surprise. In a world without nuclear weapons, the USA, with its unique, traditional military strength, would have been far more superior to any other nation. The only way for another nation to even try to defend itself against the USA would be to threaten with nuclear weapons. As long as there is one single nuclear power the least bit suspicious of the USA's intentions and political goals, it would seem that all attempts at nuclear disarmament are doomed to failure.

The only way, therefore, to eliminate the threat of nuclear weapons and the risk of a third world war is to entirely halt political violence, regardless of the type of weapons involved. This, in turn, would require total disarmament and at the same time the establishment of a global legal order, containing a system for peaceful solution of disputes between nations and ethnic groups.

Enormous savings with general disarmament

The likelihood of the realization of such a "utopia" is discussed in Chapter 6. Here, it should only be pointed out that sacrifices needed to eliminate risks of war with weapons of mass destruction are obviously different from the necessarey sacrifices connected to humanity's other mega-problems. The price – especially in the industrialized nations – can be high if we solve the problems of environmental damage and poverty, and especially high when it comes to climate change. But to eliminate the risk of the worst results of political violence does not require any economic sacrifices. Just the opposite! If we can succeed in creating a global legal order the world community would gain enormous savings, at least an estimated 1,000 billion dollars annually. Everyone would be a winner – even financially.

"Disarmament is like a party. Nobody wants to arrive before all the others."

CHANGING TIMES

SPECIFIC RISK ASSESSMENT – POVERTY

Poverty, as discussed in the previous chapter, is also linked to large potential damages, but these are still not of the same dimension as the other threats.

The probability of these damages occurring appears to be rather great and therefore risks must be considered great. But in contrast to mankind's other challenges, the problems already being caused by poverty – billions of people suffering and dying prematurely – weigh heavier than the risks of the future.

The greatest barrier toward eliminating poverty and its risks is not economic but political. Therefore, it is difficult, or perhaps even impossible, to identify the sacrifices required for corrective measures. But in this case, we can clearly state that the price can hardly be too high for necessary and suitable measures.

SUMMARY

The combined results of this discussion of the risks facing mankind are not at all encouraging. We have gotten into a very dangerous situation. We find ourselves in a risk zone for catastrophes of types and dimensions that mankind has never

Political violence affects millions of people each year. This photo shows refugees from the civil war in Liberia in the early 2000s. (Photo: Gamma/IBL Bildbyrå)

faced before. The problems have developed over short and long periods, but none came on suddenly. We could have solved the problems earlier, when they were smaller, or at least limited them through suitable measures. But instead, we let them grow so that today they create serious risks that threaten mankind's basic interests. How and why have we gotten into this situation? We will attempt to give some answers to these questions in the following chapter.

> ***"One cannot deceive someone as easily as one's self, since we like to believe what we want to."***
>
> DEMOSTHENES

4. Three invisible explanatory factors

"Two things are infinite:
the universe and human stupidity,
and I'm not sure about the universe."
ALBERT EINSTEIN

There are, of course, many explanations of how humanity has gotten into today's extremely vulnerable predicament. There have been numerous books written on this subject. But here in this section, we will concentrate on three underlying explanatory factors; three circumstances that can be seen as both the main explanation for how the present problems originated and which, at the same time form the greatest obstacles in warding off the risks and solving the problems.

How we succeed in overcoming these "circumstances" can be decisive for the future of mankind.

The circumstances that we speak of are the following:

- The continuing growth of population.
- The world's national states are far too sovereign.
- Inadequate problem insight, faulty reasoning and misjudgement by the public and politicians.

These circumstances are connected with each other to a great extent. In addition, they have one thing in common: They are not given high priority on the political agenda – if they are on the agenda at all – and they are seldom taken up in the public debate. Therefore, let's see how these factors influence the human situation when it comes to the greatest challenges of our time.

POPULATION EXPLOSION – THE PROBLEM THAT IS KEPT SECRET

As has been previously mentioned, the earth's population is currently growing by 1.2 percent annually, an increase which, if it continues, will double the number of people in two generations – or in 58 years to be exact.

The population explosion is one of the main reasons behind climate change and it also accelerates the process. For example, growing population increases demands for energy and food production, which means greater need for farming land and pastures, resulting in deforestation, which causes increased greenhouse gas emissions.

Other environmental damage, the second of our mega-problems, is also affected by population growth. For example, when population increases there is automatically less per person of already over-utilized renewable natural resources. There is the threat of famine, and more severe water shortages. Other consequences of greater population will be greater pollution, more damage to nature and faster depletion of biodiversity.

In addition, a sharp increase in population can transform normally useful and desirable advances into negative or dangerous changes. This could mean that technological developments and improved living standards – both of which are generally viewed as most advantageous – could increase demand for more energy and natural resources. If these normally positive developments occur simultaneously with a great increase in population the consequences can obviously be serious environmental damage.

Sharp population growth also increases hostility between ethnic groups and nations, as they clash over diminished resources. This would contribute to the weakening of political stability and lead to greater risk of political violence. The genocide in Rwanda in 1994 is a terrifying example.

The final point in this brief inventory: the population explosion is the most important cause of poverty. Poverty and population growth together create a vicious cycle. A small family farm can possibly support one family, but when land is inherited and divided up by several children, each with his own family, it

Drinking water is crucial for human survival and well-being. Unfortunately it is unevenly distributed geographically and is in short supply for about 2.5 billion people. (Photo: Scanpix)

can no longer support everyone. The result is worse poverty in the farmlands or in shanty-towns outside large cities. Measures against poverty are offset by greater population, which, in some nations – especially in southern Africa – result in a decrease in the percentage of poor while the actual numbers of the poor continue to increase.

It should be emphasized that most problems do not grow proportionally with the increase of population but much faster. For example, if there is a 10 percent water shortage in an area to start with – that is, water supply is 10 percent below that which is needed – then a 50 percent increase in population would mean the water shortage climbs by 500 percent to 60 percent.

What is the reason behind the obviously detrimental population explosion and why have we not taken serious measures to stop the growth (with the exception of the actions taken in China)?

One of the most important reasons for the rapid increase in population in poor nations seems to be poverty in itself, with its resultant low education level and non-existent family planning, due to ignorance and/or traditional and sometimes religious reasons.

Lack of equality between the sexes in many developing nations can also greatly contribute to high birth rates. Many young women become pregnant against their will, and parents may place greater value on sons than daughters, with births continuing until the desired number of sons is reached. High unemployment rates among women in developing nations also contribute to high birth rates.

At the start of this discussion, we noted that three closely-linked critical explanatory factors are behind mankind's four mega-problems. The population explosion – the invisible, rarely-discussed basic villain in the drama – illustrates this thesis. Lack of understanding of the problems, faulty deliberations and flawed judgment are the main reasons why political leaders have never seriously attempted to halt the extremely rapid growth of population in a number of nations, despite the fact that it has created many severe problems.

To be more specific, we can point to some plausible reasons:

- The public underestimates the significance of population growth. Some even view it as positive. Some see the increased number of people primarily as a growing market for their products.
- Many believe it is not politically correct to demand and take measures to limit population growth. Some seem to believe it is a human right to bring as many children as possible into the world.
- Politicians seem to consider population growth as a natural phenomenon that mankind should expect and that nothing can be done about it.

Even the present world political system with sovereign national states contributes to population growth since the system can effectively stop any outside attempts to influence growth. All nations today can insist that population developments and high birth rates are matters of internal state affairs.

Obviously, nobody knows today how many people can live on the earth and enjoy decent living standards in harmony with the environment. This depends, among other things, on developments in agricultural technology, future energy and water supplies, what is meant by "decent living standards", and, not the least, on the effects of climate change. But we do know that based on existing knowledge and technology the earth cannot in the long term support today's population enjoying standards that the "wealthy" nations' inhabitants are accustomed to and which all others try to attain. This means that the only rational development would be to try to adjust the number of people in relation to existing and foreseeable opportunities. Allowing wishful thinking or being resigned to the situation to guide decisions in this question means inexcusable risk-taking.

If allowed to continue, the population explosion can very well be a main cause for mankind's devastation by enormous catastrophes.

EXCESSIVELY SOVEREIGN NATIONAL STATES – AN ARCHAIC SYSTEM

The world's far too sovereign national states, the second overall explanatory factor for the greatest challenges of our time, could also be called "the lack of a global legal system." In reality, these are two faces of the same coin. The existing world political system is archaic and functions poorly in today's greatly changed and increasingly global society.

In the absence of a supranational decision-making body, we try to solve today's problems through voluntary agreements between nations. Unfortunately, this system has major weaknesses:

- Negotiations drag on. And during that time problems and risks generally worsen.
- Results are too often unsatisfactory since national interests control the negotiations.
- A great possibility that some nations do not keep their promises since the system lacks sanctions.

> ***"Humanity's salvation is only attainable when one does everything for the interest of all."***
>
> ALEXANDER SOLZJENITSYN

These weaknesses in the system worsen all of our global mega-problems, but the national states' excessive sovereignty also creates specific problem disadvantages. For example, most environmentally damaging activities are formally considered as internal state matters – such as deforestation and carbon dioxide emissions, which accelerate climate change – despite the fact that they concern us all.

Sovereign states' inability to halt political violence is painfully obvious. Not even the proliferation of weapons of mass destruction can be prevented without a global legal system.

The sovereignty of nations has also been a barrier to solving the problems of poverty. Protectionist subsidies and trade policies that have been followed for a long time, and which are still practiced by certain industrialized nations, appear to be unstoppable without a global legal system. Another consequence of the excessive sovereignty is that international aid to poor nations is often poorly coordinated and not infrequently guided by the self-interests of donor countries.

To adjust the global political system to today's realities is a far more complicated and time-consuming project than halting rapid population growth. But it is just as important.

INADEQUATE PROBLEM INSIGHT, – A BARRIER TO NECESSARY MEASURES

Inadequate insight into problems and other shortcomings is the third underlying explanatory factor for the greatest challenges of our time. This factor is invisible, mainly hiding within our own consciousness. But there is no question that it greatly contributes to all of our four mega-problems.

Inadequate knowledge of the ecosystem makes it difficult to foresee the consequences of our activities – and when we finally do see the results we may already be in the risk zone. As is the case with climate change.

Inadequate crisis awareness, like misjudgment or underestimation of problems and risks, usually results from insufficient knowledge, which, in turn, is due to lack of interest in the subject. But it can also be due to a toning down of the greatest risks by experts and politicians. One reason for that can be that politicians do not like to recognize the full extent of a risk that they cannot ward off, at least not without unpopular measures that they don't believe they can carry out. Another reason can be that politicians and/or experts do not want to alarm the public so that it becomes paralyzed and thus even fewer measures are taken.

Underestimating the risks of nuclear war is probably largely due to the fact that nuclear weapons have not been used in the six decades since World War II. This has probably lulled people into a false sense of security. But even during the Cold War's balance of terror between the USA and the Soviet Union, a third world

The Aral Sea, once the world's fourth largest lake, has lost four-fifths of its water. Human attempts to redirect nature resulted in gigantic environmental damage. (Photo: Scanpix)

war was very close to breaking out. (See section titled, "*Twice at the brink*" page 59.)

Wishful thinking is in the same category. A special form of wishful thinking is the belief that politicians can solve all problems. Another form of self deception, and possibly the most common, is the firm belief that terrible things cannot happen to me and my children.

Faulty reasoning can cause major damage and create dangerous risks. For example, short-term and nationalistic decision making instead of long-term and global reflection. Another example is faith in unrealistic or highly doubtful solutions, such as belief that economic growth or fast technological development will solve climate and poverty problems. Economic calculations that ignore possible damage to the ecosystem are another form of faulty reasoning.

And finally, it is the general public's understanding of the mega-problems, and their closely connected risks, that play the most important role in the conclusion of this global drama.

The table on the next pages, titled: *The new world community's greatest problems and risks – correlations and causes* is an attempt to illustrate the closely connected mega-problems of the new world community.

> ***"The greatest difficulty is not to get people to accept new ideas, but to get them to abandon the old."***
>
> JOHN M. KEYNES

THE NEW WORLD COMMUNITY´S GREATE

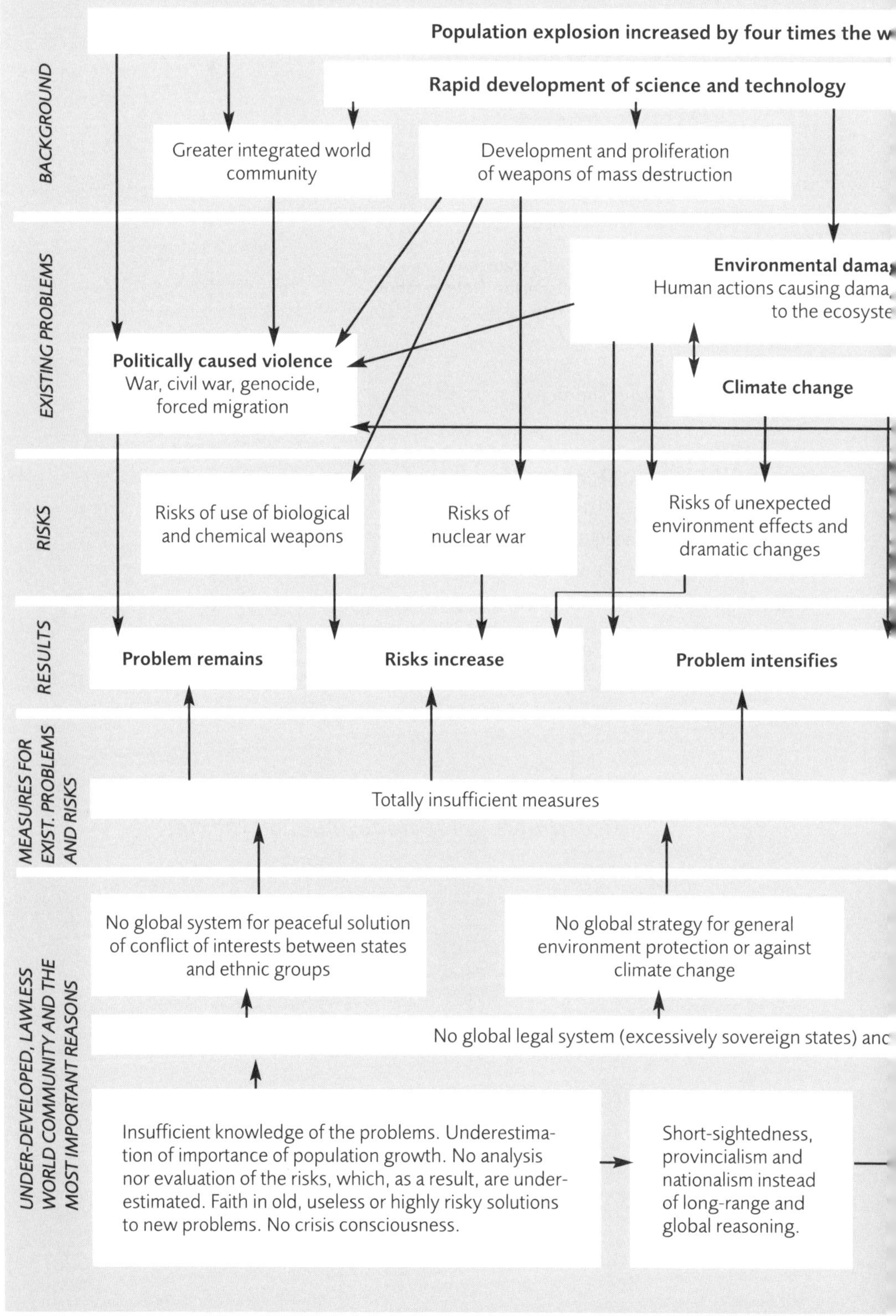

OBLEMS AND RISKS – CORRELATIONS AND CAUSES

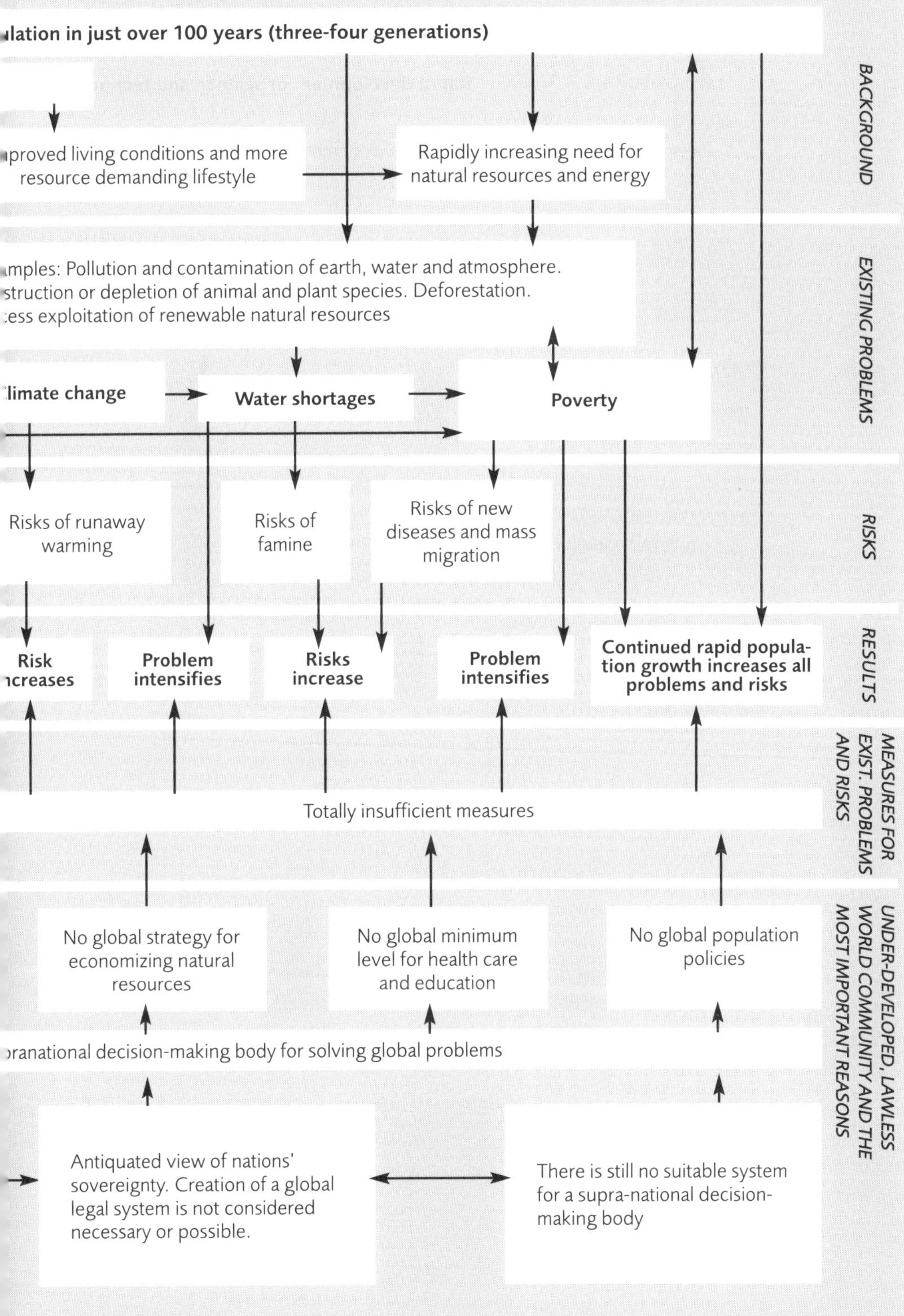

5. Difficult but necessary measures

> ***"By attempting the impossible, one attains the highest level of the possible."***
>
> AUGUST STRINDBERG

Based on current research and with the assistance of risk assessment, we have come to the conclusion that the four mega-problems can be catastrophic for large parts of mankind, if not all mankind. And when it comes to the climate threat and political violence we must agree that mankind is already in a dangerous risk zone – and that those risks increase day by day.

In order for mankind to emerge safe and sound from these problems we must not only take suitable measures but also have a bit of luck. And the later the measures are taken, the greater role luck will play in mankind's future. Right now, the outlook is everything but bright because:

- global problems can only be solved through global action,
- but global action requires global decisions,
- and global decisions can only be made by a supranational body,
- but no effective, supranational, decision-making body exists today.

It would be ideal if the world's political leaders could succeed in creating a supranational, generally-accepted decision-making body for global questions. But this would first require having some kind of global legal system. However, creating such a global legal system would take many years, even if there were a desire by all national states to do so – and there is obviously no general consensus whatsoever for this today.

The threats that mankind faces, however, must be tackled immediately since all delays in action programs will result in increased risk. Political leaders must therefore compromise and continue to negotiate and agree on the most rational measures possible, under today's inefficient and outdated decision-making processes. At the same time they must also work toward long-term, sustainable solutions to problems by devoting significant resources to create a global legal system as quickly as possible. Until then, such work must be carried out in parallel on several fronts to prevent – if possible – making irreparable mistakes.

Measures can be taken either to solve specific problems or to handle the invisible explanatory factors discussed in the previous chapter.

MEASURES FOR SPECIFIC PROBLEMS

When it comes to *climate change*, not only Nicolas Stern, author of the respected Stern Review, but practically all experts and specialists now stress the need for a common global vision, multilateral framework and coordinated action. These are the decisive factors for success. However, what's missing is a practical directive for how such productive and absolutely necessary cooperation can be obtained within the framework of the existing world political system.

A key question, therefore, is how we can increase the chances that traditional climate negotiations will lead to more rational decisions and more effective measures than have been attained so far. I believe it is worthwhile to test two possible actions.

The first would be for political leaders to ask scientists to make a risk analysis of the main issues. That is, to investigate and report on the worst damages that climate change may potentially cause and at the same time estimate the probabilities that these damages will occur. I do not really believe that the results of such analyses, made by specialists and based on all available scientific evidence and knowledge, will deviate to any great extent from my previously noted sketchy analysis of the dimensions of risks.

Population growth is the main reason for poverty. Poverty and population growth combine to create a vicious circle. (Photo: Scanpix)

However, such professional risk analysis would attain far greater credibility with both politicians and the public. And credibility is of decisive significance for possible political acceptance of long-range and perhaps painful measures to lessen the climate threat.

The second action would be to have studies and thorough discussions – carried out by the public in respective nations and between nations – of some fundamental questions. This would be of great practical importance for both international negotiations and for the public's attitude toward climate measures. At least three questions should be thoroughly debated:

1. How should the goal for climate policy be truly defined?
2. What sacrifices would be rational to limit climate risks?
3. How should the costs (the sacrifices) for the climate measures be fairly shared between nations?

The three questions are naturally linked together, but shall be discussed separately for the sake of clarity.

The goal for climate policy should be studied extra carefully. In recent years, politicians have found a simple but dangerous answer to the question. This is that global warming should not exceed a certain limit, currently 2 degrees C. In this way, it is expected that mankind will avoid being hurt by devastating climate catastrophes.

But such a goal is actually unsuitable for four reasons, of which three were noted in the risk analysis, but are worth repeating:

- We do not know what temperature increase will lead to disastrous climate catastrophes. As previously noted, there is great uncertainty in this question. The probability is far from negligible that a critical increase can be 2 degrees C above the historic level, but it could also be at a lower level.
- We do not know which greenhouse gas concentration will lead to a specific temperature increase.
- We do not know how much emission (the only factor that we can influence) will result in a specific greenhouse gas concentration.

A fourth reason for not setting precise climate goals of this type is that it would lull the public into a sense of false security. It is a grave mistake to believe that no climate catastrophe can occur if temperatures increase a maximum of 2 degrees C, or even be kept below 2 degrees C.

In order to determine the highest permissible emissions, with the goal of preventing mega-catastrophes, we are forced to trust some kind of average value, based on obviously uncertain estimates in each of the three stages. It is highly regrettable, to say the least, to allow the solution to this central question for mankind's future to be based on an accumulation of errors and uncalculated risks. This demands a different approach.

Considering that we humans can no longer eliminate the increasing risk of climate catastrophes and can only reduce the rate of increase, the obvious goal would be to minimize the risk. In practice, this should mean that we reduce emissions as much as possible – but on the condition that in minimizing the risk of catastrophes we do not sacrifice our interests in the process. Identifying those interests would certainly require long, tough debates, but if this process provides the public with deeper knowledge of the values that are threatened and what the measures would probably cost, this would make possible significantly more effective action. In other words: much less risk of serious catastrophes.

And with that, we have moved on to the next question.

How much sacrifice is rational? The question is really the same as the last step in a dependable risk assessment and should thus be discussed based on a previously completed risk analysis.

But whether a risk assessment has been made or not, and independent of how the goal for climate policy measures is formulated, the question of rational sacrifices has several moral dimensions. This can partly be discussed and decided on the basis of common sense and available facts and knowledge. It is also partly a question of values that can not be decided on objective grounds and therefore does not have any obvious "correct" answer.

The public's understanding of the scope of the potential damage, the likelihood of probability and the value of what can be sacrificed, can obviously differ and the differences can depend on several factors. For example, we have varying aversion to risk; we take different views of whether potential damages can occur in only 50 years or in 500 years; and we can feel that reducing material living standards would be negative at various levels. Even age plays a role. A teen-ager, for example, runs a far greater risk of being personally harmed by climate change and should therefore be more willing to pay for various climate measures than an 80-year-old, who could only be affected by having to pay for some of the costs. Obviously, people are also motivated to make sacrifices to reduce dangers based on whether their own children and grandchildren are threatened, or whether anonymous future generations are threatened.

Thus we cannot expect any general agreement on such value judgments, not even among people in the same nation. Nevertheless, it is obviously important that the public is given the best possible background and knowledge for a well-founded understanding of both the factual and value judgment sides of the question. This usually increases the probability for achieving a majority opinion and even to win acceptance for necessary, weighty political decisions.

It will be even more difficult to reach some consensus between nations on the question of the size of sacrifices that are rational in order to limit climate risks. Some firm idea of the total cost will most probably be greatly affected by answers to the next question: how costs should be fairly shared.

"Judging from the main portions of the history of the world, so far, justice is always in jeopardy."

WALT WHITMAN

The moral responsibilities of the West

When it comes to *the question of how to fairly share the burden of sacrifice* for the imminent climate measures, the industrial nations' moral responsibilities must be considered. The main problem is how. Until now, the rich nations have contributed by far the most to the increase of carbon dioxide in the atmosphere. The difference in emissions per capita between the rich and the poor nations are extremely large even today. At the same time, the poor nations generally have been harmed, are being harmed, and will be harmed faster and more seriously by global warming. Wise political decisions on this question must be based on long and intensive debate in the rich nations. But even then we cannot expect any unanimity, neither within nations nor among nations.

In the best case scenario, greater knowledge of the increased risks of climate change will result in creating, in the relatively near future, a supranational, decision-making organization devoted solely to the climate question.

If this does not succeed, and if humanity does not have the help of good luck from unknown ecosystem effects or unexpected technological innovations, then there are probably only two scenarios that can get politicians to unite on a rational and effective global climate program. One is that the world's public – perhaps youth in particular – gains a much deeper knowledge of the risks associated with climate change and forces leaders to change their strategy. The other is that a major catastrophe, caused unquestionably by climate change, occurs and opens all eyes to the threats.

Industrial nations' lifestyle in focus

Otherwise, the climate front is not at all bright in the short and medium term perspectives. There were few, if any, solid decisions that emerged from the large international climate meeting in Copenhagen in December 2009. And the outlook for results from future negotiations appears dim. In order to reach unanimity even on insufficient, yet important, measures, the industrial nations must be prepared for immense "costs" – and in the long term probably even sacrifices in the form of changed lifestyle. This can

Current dilemma: Coal-fired power plants are the industrial nations' main supplier of electricity, but also one of the largest sources of environmental damage because of their huge emissions of carbon dioxide. Photo shows the Jaenschwalde coal-fired power plant in Germany. (Photo: Sean Gallup/All Over Press Sweden)

be a really crucial problem. The industrial nations' negotiating position is highly unfavorable. Expectations that we will be able to combine effective climate measures with continued economic growth of the old pattern are clearly not only over-optimistic but even naïve, at least in the medium term perspective. A rapid switch to "clean" sources of energy would require enormous expenditures. Investments in energy and climate research would have to be increased many times over current levels. A fair and effective global package of measures to solve climate issues would require enormous resource transfers from rich to poor nations, either financial subsidies to developing nations' climate measures and/or payments for so-called emission rights. No matter how optimistic we are, it's obvious that the equation would not balance without painful reduction of consumption in the industrialized nations.

In a longer perspective, the economic calculations become more hopeful. The tremendous capital transfers can, at best, be a decisive factor in getting the poorest nations on their feet. Thus, not only would the enormous income difference between the industrialized and developing nations be reduced but the global economy would also gain vast human resources, resources that have so far been wasted.

Even when it comes to humanity's second mega-problem, *environmental damage* (excluding climate change), all nations would have to agree to a binding set of global rules and regulations in order to bring about effective measures. However, the question is whether it is possible to stop many of the activities that damage the environment as long as environmental questions are formally considered as internal, domestic affairs of each sovereign state.

Under existing circumstances, political leaders hope that multilateral agreements will slow down the most damaging types of environmentally damaging activities, such as deforestation and fish stock depletion. In order to alleviate environment problems in the poor nations, it can be necessary for industrial nations to provide economic compensation and in this way "pay ransom" to save the environment.

Another desirable measure, which should be quite easy to gain consensus on, is that environmental consequences will be measured in economic calculations in an entirely different way than they have been to this point. (See side section, titled "Economic growth and environment – Misleading pricing", page 86.)

Ranking high on the "wish list" are greatly intensified research on the ecosystem and necessary assistance measures to improve water supplies in many developing nations. But the real problem here, as in many other problem areas, is how costs will be shared by the world community.

When it comes to *political violence*, we can hardly imagine any permanent solution that can eliminate risks connected with weapons of mass destruction without creating a global legal system. While waiting for such a global system, political leaders will have to test all partial solutions and adjustments that can reduce violence and risks.

The poverty problem can hardly be solved before we have successfully halted the rapid increase of population in the poorest nations. But programs for this should be far more effective than they are today. For example, assistance can be more effective by coordinating efforts within the framework of the UN. A number of aid programs are still controlled more by the donor nations' own internal interests than the needs of the recipient nations. A common strategy by donors toward rogue nations and thoroughly corrupt leaders would also mean much for the people suffering in those countries.

"There has never been a good war or a bad peace."

BENJAMIN FRANKLIN

Misleading pricing

Everyone desires the best possible quality of life. Most try to acheive it by improving their material standards. Historically, there has been a clear connection between increased Gross National Product, GNP, and higher material standards of living. This presumably explains why just about every economist and politician extols and makes every effort to increase the GNP. But we seem to forget that the GNP provides erroneous signals when natural resources are excessively utilized – and no signals at all when we humans damage the ecosystem. As a result, GNP growth itself can be harmful if it is attained by damaging our most important interests.

Naturally, we can try to influence this development by pointing out shortcomings in national economic accounting. For example, we can urge business and industry leaders to include the damages to the ecosystem that could result from their financial decisions. Unfortunately, neither economists nor business people can do much about the problem. The market economy is controlled by factors expressed monetarily, factors that are an expense or provide income. And the market economy's efficiency is insured by competition among the players. Therefore it is futile to try to require business decision-makers to include in their calculations useful factors that lack market prices – since nature offers them free – or to refrain from damaging the ecosystem when nobody demands compensation. If responsible companies nevertheless attempt to heed moral appeals about environmental damage, they risk being hurt by competitors who do not take such considerations into account.

Thus, it is society's obligation and responsibility to save natural resources and protect the ecosystem. The task is not easy, but it is inevitable and it is obvious where the responsibility lies. With the help of scientific and economic expertise, political leaders must establish limits for how and to what extent we can exploit and utilize nature without detrimentally affecting humankind's living conditions. Political decisions should prohibit or punish actions and enterprises that are unacceptable. This naturally applies to making rules for business and industry, which happens occasionally – such as prohibiting cartels. Thus, politicians, through taxes and fees, must establish prices for natural resources (both renewable and non-renewable) whose exploitation should be limited. This means that a correct price for a product or commodity would also include costs of production and waste to ensure that nature is not damaged. Naturally, it is most important that prohibitions and taxes and fees are applied equally to all competing in all countries. Therefore, this is also a problem that must be solved on a long-term global basis.

MEASURES TO HANDLE CLOSELY LINKED PROBLEMS

In the previous chapter we described several closely linked explanatory factors of the mega-problems facing mankind. The urgency of tackling these "circumstances" is quite obvious since they often also create barriers to logical solutions of the serious overall problems.

For example, what can be done about rapid *population growth*? Indeed, the key to the solution is obviously the general public's understanding of and attitude to the problem. Therefore, in the near future, the worldwide public must be conscious not only of the dangerous consequences of the population explosion described earlier but should also know:

- that the rapid population growth is not any one nation's "internal domestic affair" but something that can harm humanity's most important interests,
- that it cannot be considered a human right for a woman to have more than two children,
- that the earth's resources are not even enough for the existing population if all were to enjoy industrialized nations' living standards – and that the situation will be much worse as early as within two generations if the world population has increased by several billion,
- that we cannot count on people in the poorest nations continuing to accept living on so much fewer resources than people of industrialized nations,
- that a "normalization" of average life expectancy to that of industrialized nations would in itself mean that the number of people on earth would increase from today's 6.8 billion to 8.0 billion.

It is unrealistic to believe that such an educational campaign would be undertaken by political initiative, given the nonchalance with which this problem has been treated so far. Instead, the public must first be clearly aware of the great risks connected to the population explosion and demand counter-measures before the matter finally gets priority on the political leaders' international agenda.

Only then can we hope for an action program worth the name; that negotiators from the rich nations agree with colleagues from the fast-growing nations to adopt the most effective efforts in education, improved health care, greater employment for women, family planning and other measures to curb the birth rate.

Information as powering change

The first step toward solving a problem is usually realizing that there is a problem.

At the end of the previous chapter, and as is shown in the analysis of population growth, we stated that the *public's lack of knowledge* of the great global threats is – both directly and indirectly – one of the greatest barriers to long-term and effective solutions of these problems.

This is easy to understand. But it is extremely difficult to take preventive measures quickly enough, especially when the problems and risks steadily increase. Researchers, scientists, educators at all levels, journalists in all types of media, bloggers and all others in discussions on the Internet or out among the public have an extraordinarily important task before them.

But it is political leaders who have the main responsibility to provide the public with relevant and easily understood information on all the important problems and risks that threaten mankind. A well-functioning democracy depends on the public majority's having well-founded understanding of the most important problems, especially when rational, long-term solutions to these problems require uncomfortable measures. Otherwise, governments and authorities will either make unsatisfactory decisions – perhaps even decisions that can lead to catastrophes – or also risk nations being taken over by irrational politicians or power-hungry populists.

It is well-known that a democratic political system is no guarantee that correct political decisions will always be made. The system's weakest point is the dependence on the quality of the majority opinion in a critical situation. We cannot sit back comfortably in the knowledge that we have the world's best political system – and remain blind to the system's shortcomings and risks. We must instead increase understanding and awareness of the risks and try to eliminate them, or, if this is impossible, to reduce them as much as we can. We must strengthen as much as possible the public's understanding and basic knowledge of the most important questions facing society. Perhaps the most effective way to do this is to make the public debate of these questions more vigorous, more structured and goal oriented, and in doing so make the public discussion more interesting, popular and rewarding.

Risk consciousness is also necessary and valuable when there are important, real risks. This does not paralyze the public but rather prepares people to take all measures possible to save themselves from peril. Risk consciousness can actually create a decisive condition for the public's accepting and taking necessary measures against threatening catastrophes. In this connection, there are major shortcomings in communications connected with climate change. A more meaningful discussion on the risks of various strategies for emission reductions would increase consciousness about the problem and thereby probably also willingness to accept stronger measures.

Parallel to the information efforts and parallel to all important negotiations and decisions regarding the greatest global problems that are carried on by current methods, a high-priority effort must be undertaken to seriously start lengthy and difficult procedures to reduce the sovereignty of national states to an acceptable, harmless level. Or, in other words, to create a global legal system.

The following chapter discusses what such a new world order would mean, and whether it is possible in the foreseeable future to reach the goals of such a project.

"The ability to think today different from yesterday distinguishes the wise from the stubborn."

JOHN STEINBECK

6. Global legal system – sooner or later

"We can't solve problems by using the same kind of thinking we used when we created them."

ALBERT EINSTEIN

That the world must have a functioning global legal order is not at all a new concept. Following World War I, for example, the League of Nations was formed. But it was a failure. And after World War II, the United Nations, the UN was created and it too failed to gain the decision-making power or authority to attain its objectives.

Therefore, an obvious central question right now is whether mankind, in the foreseeable future, will be able to succeed in creating a global legal order worth its name.

I do not know, but I do know it is necessary.

I also believe that it will occur, sooner or later, and hopefully before it is too late to produce a global, effective action plan to handle today's major global challenges, an action plan that is respected and followed by all involved nations.

I am also certain that those political leaders who succeed in implementing a functioning global legal order will be ranked as the most important statesmen in history.

Today, we all live in a global world community – whether we like it or not. Thus people of all nations can be influenced by how people of all other nations behave – and not only when it concerns our mega-problems. For example, it's estimated that the latest financial crisis is contributing to the deaths each year of some 300,000 more children in poor nations because of reduced foreign aid.

Deforestation in the poor nations can threaten the effects of

The UN today cannot handle the challenges facing the world community. Therefore, the organization must be reformed from its very foundations. (Photo: IBL Bildbyrå)

climate policy measures. Continued population growth can lead to food shortages, greater climate crises, international conflicts and much more.

National states' unlimited sovereignty stands out today as both antiquated and precarious. It is hardly surprising that the new, still greatly undeveloped, global community suffers from the typical disorders of developing nations, such as political instability and the lack or non-existence of institutions. Consequently, and with good reason, we can argue that the new emerging world community is still uncivilized, incompetent and immoral.

Uncivilized – because the resolution of conflict of interests between nations is often violent, since the world lacks global laws, courts of justice and law enforcement agencies.

Incompetent – because there are no decision-making organizations to solve humanity's global problems.

Immoral – because the world community allows a large part of humanity to die of poverty or live in extreme misery.

Who would choose to live in a national state that looks like our world community? People in the USA, for example, would live in a nation where brute force or economic power decides legal disputes, where more than one in ten people would be chronically undernourished because of poverty, and where 400,000 American children under five would die each year of the same cause.

Nationally, we live in communities governed by law, but globally we live in (or near enough to) a lawless society.

Necessary building blocks

What do we mean by a global legal order? First, it must be emphasized that it is definitely not a matter of attempting to create some type of united states of the world, like the USA. This is impractical because there are simply too many and too great differences among nations' languages, cultures, economies, and political systems. The goal should be to create as soon as possible a supranational decision-making body that can overcome mankind's common problems, and that can form the basis for a world society that is not only incomparably more secure than the present system but is more like most peoples' ideal society.

The following building blocks are necessary:

- A neutral, supranational decision-making assembly for global and international questions. This would require a "watertight" system that would guarantee that no state or group of states could use the system to force decisions that favor their own interests at the cost of other nations or mankind.
  In other words, extremely well-balanced principles would be required for composing this assembly. Combined with suitable rules for the decision-making process, these principles would prevent abuse.
- Conflicts between nations (and in some cases between ethnic groups) must be decided by international courts. A neutral,

supranational law enforcement body would replace all national so-called defense forces.

In no way is the aim to eliminate national states, but only to reach a reasonable limitation of national states' sovereignty. The limits should be set so that nations' or humanity's important interests are not unjustifiably harmed.

"Questions that are settled by violence are never settled."

JAMES JOYCE

Nearly only advantages

Global development since World War II has continued to provide us with an increasing number of arguments for a new, global legal order. Let's point out some obvious advantages:

- Faster and more efficient measures to handle all global problems and risks.
- Improved, faster and more impartial solutions to disagreements between nations and ethnic groups.
- The end of mass killing, injuries and human suffering in wars and civil strife.
- Most reasons for acts of terror would disappear.
- Terrorists would face great difficulties obtaining heavy weapons or weapons of mass destruction.
- Basic human rights to worldly goods would be realized.
- The USA would no longer need to be both a superpower and world police force, and an object of hate and target of terrorism.
- No destruction, or at least greatly reduced damage, to material and cultural values in the wake of violence.
- More effective combating of international organized crime.
- Savings of 70–80 percent of the world's "defense" costs, the equivalent of 1,000 billion dollars a year.

The disadvantages of a global legal order are in theory almost non-existent. Leaders of undemocratic nations, who place their own personal interests above those of their people, would naturally be on the list of losers. That list would also include politicians who strive for special positions of power, for their nations or themselves. Such disadvantages are hardly worth considering.

Obviously, it is a common reaction when we point to the need for a supranational decision-making body as a mean to solve serious global problems, to say that such a system would (only) lead to more bureaucracy and more corruption. But this objection is not a valid argument against establishing a global legal order. We only need to remember that bureaucracies are found in all state activities. And so is corruption, although to an even more various degree. Yet, no one proposes that all state authorities should be abolished in order to eliminate bureaucracy and corruption. The services that these authorities provide are obviously extremely important, if not essential. For the public in most national states, the value of laws, safety, courts, general health, functioning infrastructure, and much more – in brief, the foundation for a just society – is incomparably greater than society's costs for bureaucracy and corruption. It would be the same in a new world society if a global legal order were created. The previously-listed advantages of such a system should be sufficient basis for this conclusion.

Thus, it is very easy to see the necessity and advantages of a global legal order. But, unfortunately, experience shows that efforts to achieve it are far more difficult in practice

Mutually devastating suspicion

There are difficult obstacles to a new global legal order, for example:

- Emotional and ideological resistance to "supreme authority".
- Nationalism.
- Hostility and suspicion, mutually or one-sided, between many nations.
- The UN's functional obstruction, organizational shortcomings and insufficient prestige – despite the obvious route to a global legal order should logically go through a basically reformed UN.

Without a doubt, these are very difficult barriers to overcome, and we could easily make the list longer. But none of the barriers are impossible to overcome and none create a solid argument against a global legal order. The most difficult, real obstruction is that presumably there is not yet a detailed proposal for the system described above, a secure system for a supranational, decision-making body.

However, it would be unwise and highly unfortunate if all of these obstructions to a global legal order caused the project to be written off as an unrealistic pipe dream.

The conclusion would be different when we fully realize:

- that the threats to mankind are extremely great,
- that the probabilities of very large catastrophes are not negligible,
- that the risks continually increase, and
- that the existing political system of totally sovereign states is incapable of solving these problems or minimizing these risks.

The only rational conclusion must be that we should work toward changing over to a system that can deal with the problems. Naturally, we will continue to face powerful objections and great obstructions to a system change. For example, that there is no detailed, concrete proposal for a better system. But should we accept the situation, sit back, and simply hope for luck, that is, that catastrophes will not occur? Or should we attempt to tear down the obstructions and do everything we can to contribute to creating a competent system?

Here, it is also important that the new system does not require that all nations must be democratic in the western sense of the word. Such a requirement would make meaningful negotiations impossible from the very start – consider, for example, the new industrial giant, China – and thereby would irresponsibly increase the risks of really serious catastrophes for mankind.

Instead, the first requirement is that political leaders work for

The world's top political leaders must show extraordinary qualities of statesmanship and willingn
to compromise. But do they have the ability? Do they understand the gravity of today's challenge
From left: China President Hu Jintao; German Chancellor Angela Merkel; Russian Prime Minis

'ladimir Putin; Brazilian President Lula da Silva; USA President Barack Obama; and Indian Prime
Minister Manhan Singh. (Photo: Scanpix and IBL Bildbyrå)

their peoples' long-range welfare. It is quite obvious that creation of a global legal system is in the interest of all nations. Just as a healthy body organ can not function for a long time if the body is attacked by a deadly disease, a very healthy national state cannot remain healthy for a long time in a sick world community.

The decisive task for policymakers must thus be to design a watertight system, as described above, which would prevent abuse by the supranational control body. But what is the solution if we do not succeed in designing a system that entirely eliminates abuse? Well, then we must carefully analyze and assess the risks of the system that is closest to the ideal for supranational decision-making, and compare these risks with those that are linked to retaining the political system of fully sovereign nations. If we then find that the risks with the new system are obviously smaller we should adopt it. Otherwise, the old system will surely remain and the vision of a global legal order will be only a vision for the time being.

In conclusion, we can state that nationalism has served its purpose over a long time, and it has now played out its role. Solidarity (mutual consideration of each other's interests) among members of a group first extended out from a family to a tribe. In the next historic phase the same thing took place, but from the tribe to a region and then to a nation. This change was not painless but it was necessary for survival. Now, the final step must be taken to global mutual consideration for the same reason, but today it does not mean only survival for a family and tribe but nations and – truly – the human species itself.

"Nationalism is your conviction that your country is superior to all other countries because you were born in it."

GEORGE BERNARD SHAW (REVISED)

7. Summing up

"A politician thinks of the next election,
a statesman of the next generation".
JAMES F. CLARKE

Some may surely think that the views of the world situation as presented in this book are far too pessimistic. But I maintain that they are, if anything, realistic.

I will not deny that there are chances that with luck and skill we can solve the problems and escape the threats relatively unscathed, perhaps even without making extraordinary efforts. It's not impossible that researchers will soon discover biotechnological solutions that insure humanity's food supplies at an acceptable cost and without harming the ecosystem – even if the world population climbs to 10 billion or more in two generations. Nor is it impossible that scientists will quickly succeed in developing cheap and highly efficient solar energy techniques that solve the global energy problem without creating greenhouse gases that damage the environment. And, naturally, it is not entirely out of the question that we humans have such great luck that runaway warming does not begin and that no weapons of mass destruction are ever used.

Everything is possible. But in our situation is it wise to put our trust in highly uncertain eventualities and good luck?

The main rule for all who work in high risk businesses, and who want to survive, has always been to never gamble more than you can afford to lose. This rule applies just as well to gambling as to business and politics. Today, all of mankind's future is in the risk zone.

In the foreword to this book, the financial crisis is referred to as having emerged from America's credit-financed, large scale consumption, consumption that was based on over-optimistic expectations. Counting your chickens before they are hatched has always meant taking a risk. We have been and are continuing to repeat the same type of mistake on an unparalleled greater scale.

We allow the earth's population to increase at a rapid rate without knowing how it will be provided for.

We don't even try to reform the system of having totally sovereign states, despite an increasing number of states possessing weapons of mass destruction.

We allow carbon dioxide in the atmosphere to increase more than necessary, despite the fact that this can damage mankind's basic living conditions.

We do all this in the hope that that luck and new technological achievements will solve the problems, and perhaps also in the expectation that the world's politicians suddenly begin to prioritize humanity's long-range interests ahead of their own nations' short-term interests.

In other words, we humans today recklessly violate the aforementioned rule of risk, and play a game of chance, putting up a stake that we definitely can't afford to lose.

What the world needs most today are sovereign statesmen in less sovereign states.

In our situation, is it wise to trust highly uncertain eventualities and luck?
(Photo: Scanpix)

POPULATION DEVELOPMENT OVER 2000 YEARS

What are the limits of sustainable development? The earth's population since the start of industrialization has increased 8.5 times, from 0.8 billion to 6.8 billion. The UN forecasts that the number can increase to 9-10 billion in about 2050.

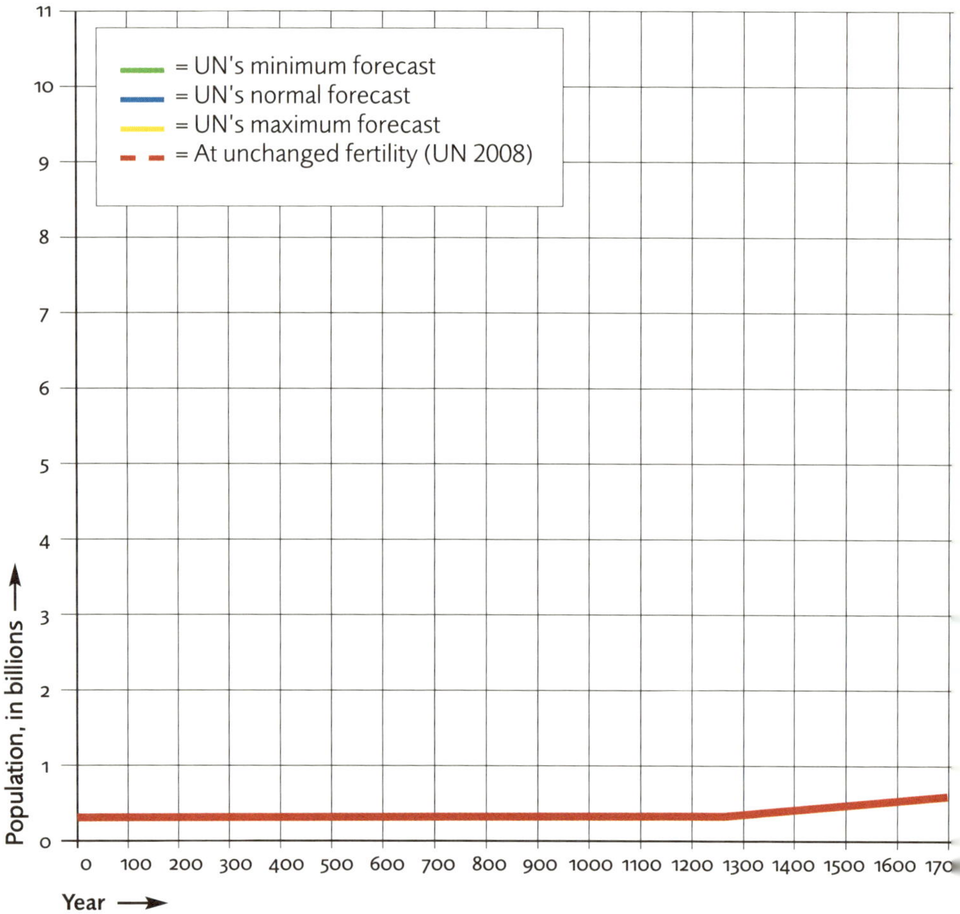

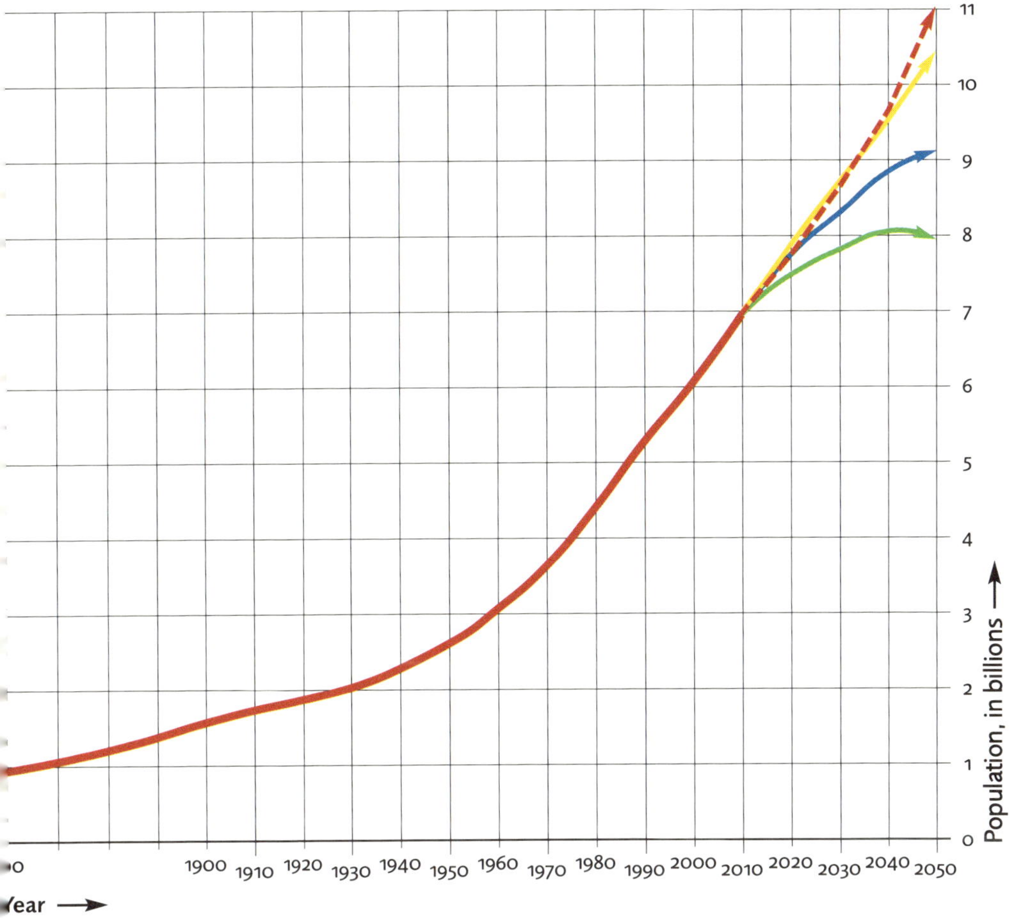
11
10
9
8
7
6
5
4
3
2
1
0
Population, in billions
1900
1910
1920
1930
1940
1950
1960
1970
1980
1990
2000
2010
2020
2030
2040
2050
ear

Almost 70% of the world's population is in 11 nations + EU

Nation/Region	Number of inhabitants 2009 millions	GNP per capita 2009 USA $	Natural population growth 2006 %	Life expectancy Female 2007 years	Life expectancy Male 2007 years	Literacy 2006 %	CO_2 emissions 2007 Million tons	CO_2 emissions 2007 Tons/capita
China	1 346	3 565	0.5	74	71	91	6 083	4.6
India	1 198	1 032	1.6	67	63	61	1 370	1.2
USA	315	46 442	0.6	81	75	98	5 854	18.9
Indonesia	230	2 223	1.3	70	67	91	389	1.7
Brazil	194	7 737	1.4	76	68	89	350	1.8
Pakistan	181	1 016	1.9	65	64	50	140	0.8
Bangladesh	162	559	1.8	66	64		42	0.3
Nigeria	155	1 089	2.4	44	44		54	0.4
Russia	141	8 873	-0.4	72	59	100	1 579	11.1
Japan	127	39 573	-0.1	86	79	100	1 235	9.6
Mexico	110	8 040	1.3	79	74	91	451	4.2
SUB-TOTAL	**4 159**	**(61 % OF THE WORLD POPULATION)**					**17 547**	
EU (27 countries)	496	32 527	0.1	82	76	100	3 971	8.1
TOTAL	**4 655**	**(68 % OF THE WORLD POPULATION)**					**21 518**	

Sources: Swedish Institute of International Affairs, August 2009.
International Energy Agency for carbon dioxide emissions.
Note: In this table, data for the number of inhabitants and GNP per capita is for 2009, one year later than the 2008 data in the following table of 195 sovereign states. Figures for carbon dioxide emissions are for 2007, which is two years later than the 2005 figures in the following table. (Tons are metric tons.)

*

Comments

As is seen in the above table, each of 11 nations has a population of more than 100 million. Combined, these most populous nations comprise 61% of mankind. If this group is increased by 14 nations (each having between 50 and 100 million inhabitants), it would account for 75% of mankind. Adding the next group of 58 nations, each with populations between 10 and 50 million, would increases the share by 20 percentage points. This means that 95% of the world's population, or 6.5 billion people, live in 83 nations. The next group, which consists of 68 nations each with population of 1 to 10 million, has only 5% of the world population. When combined, the smallest 44 nations, each with fewer than 1 million inhabitants, make up only 0.2 % of mankind.

192 nations were members of the United Nations, UN, at the end of 2009. In addition, there are three independent states that are not members for various reasons. But in the table starting on page 105, we have included these three, Kosovo, Taiwan and the Vatican State.

Brief facts on the world's 195 sovereign states

STATE	NUMBER OF INHABITANTS	GNP/CAPITA US DOLLARS	POPULATION* INCREASE/Y, %	LONGEVITY FEMALE	LONGEVITY MALE	LITERACY %	CO_2/CAPITA TONS/YEAR**
Afghanistan	28 200 000	460		48	47	28	0.02
Albania	3 200 000	4 250	0.7	77	72	99	1.1
Algeria	34 400 000	4 920	1.6	74	71	70	4.2
Andorra	66 000	38 800		85	81	100	
Angola	17 500 000	5 700		43	40	69	0.6
Antigua & Barbuda	85 600	13 390	1.1	77	73		
Argentina	39 900 000	8 520	1.0	79	72	97	3.9
Armenia	3 000 000	3 400	0.3	75	68	100	1.4
Australia	21 000 000	50 150	0.6	83	78	100	18.1
Austria	8 400 000	52 160	0.1	82	77	100	8.9
Azerbaijan	8 500 000	6 140	1.1	71	64	99	4.4
Bahamas	335 000	20 590					
Bahrain	766 000	25 245					
Bangladesh	161 300 000	510	1.8	66	64		0.3
Barbados	295 000	13 700				100	
Belarus	9 600 000	6 060	-0.6	74	63	100	6.5
Belgium	10 500 000	49 430	0.1	83	76	100	9.8
Belize	294 000	4 320					
Benin	9 300 000	860	3.0	56	55	36	0.3
Bhutan	700 000	2 090		66	64		
Bolivia	9 700 000	1 890	2.1	68	63	87	1.0
Bosnia and Herzegovina	3 900 000	4 850	0.0	77	72	97	6.9
Botswana	1 900 000	8 930	1.1	32	34	81	2.5
Brazil	194 200 000	8 680	1.4	76	68	89	1.7
Brunei	398 000	43 750					
Bulgaria	7 600 000	6 850	-0.6	76	70	99	5.7
Burkina Faso	15 200 000	590	3.1	50	48	22	0.1
Burundi	8 900 000	113	2.9	46	44	60	0.03

STATE	NUMBER OF INHABITANTS	GNP/CAPITA US DOLLARS	POPULATION* INCREASE/Y, %	LONGEVITY FEMALE	LONGEVITY MALE	LITERACY %	CO_2/CAPITA TONS/YEAR**
Cambodia	14 700 000	740	2.1	61	54	75	0.04
Cameroon	18 900 000	1 290	1.9	47	46	69	0.2
Canada	33 200 000	47 070	0.4	83	78	100	16.6
Cape Verde	542 000	3 665					
Central African Republic	4 400 000	480	1.9	40	39	49	0.1
Chad	11 100 000	935	3.3	45	43	27	0.01
Chile	16 800 000	10 810	1.1	82	76	96	4.1
China	1 336 300 000	3 180	0.5	74	71	91	4.3
Colombia	46 700 000	5 175	1.5	76	70	93	1.4
Comoros	860 000	855					
Congo, Dem. Republic of	64 700 000	210	3.2	46	44	68	0.03
Congo, Republic	3 800 000	3 660	3.2	55	52		0.6
Costa Rica	4 500 000	6 730	1.3	81	76	95	1.7
Croatia	4 600 000	14 415	-0.2	79	72	98	5.2
Cuba	11 300 000		0.3	80	77	100	2.2
Cyprus	864 000	32 195					
Czech Republic	10 200 000	21 040	0.0	79	73	100	11.7
Denmark	5 500 000	67 390	0.2	80	76	100	8.5
Djibouti	848 000	1 240					
Dominica	75 000	5 090	0.8	76	72	96	
Dominican Republic	9 900 000	5 130	1.8	72	65	87	2.0
Ecuador	13 500 000	3 930	1.7	78	72	91	2.2
Egypt	76 800 000	2 110	2.0	73	69	71	2.4
El Salvador	7 000 000	3 070	1.8	75	69	82	1.0
Equatorial Guinea	520 000	16 260					
Eritrea	5 000 000	295	3.0	58	54		0.2
Estonia	1 300 000	18 800	-0.2	78	67	100	13.5
Ethiopia	85 200 000	320	2.6	49	48		0.1
Fiji	844 000	4 315					

STATE	NUMBER OF INHABITANTS	GNP/CAPITA US DOLLARS	POPULATION* INCREASE/Y, %	LONGEVITY FEMALE	LONGEVITY MALE	LITERACY %	CO_2/CAPITA TONS/YEAR**
Finland	5 300 000	54 577	0.2	82	76	100	10.1
France	61 900 000	48 010	0.4	83	77	100	6.2
Gabon	1 400 000	10 940	1.8	54	53		1.2
Gambia	1 800 000	480	2.3	59	56		0.2
Georgia	4 400 000	3 060		75	67		1.1
Germany	82 500 000	46 500	-0.2	82	76	100	9.5
Ghana	23 900 000	790	2.2	58	57	58	0.3
Greece	11 200 000	33 430	0.0	81	76	96	8.6
Grenada	105 000	6 180				96	
Guatemala	13 700 000	2 660	2.8	72	65	69	0.9
Guinea	9 600 000	430	2.9	55	54	31	0.2
Guinea-Bissau	1 700 000	260	3.1	47	44		0.2
Guyana	736 000	1 485					
Haiti	9 800 000	790	2.1	54	53		0.2
Honduras	7 200 000	1 800	2.2	71	67	80	1.1
Hungary	10 000 000	16 340	-0.3	78	70		5.6
Iceland	303 000	60 120	0.8	82	77	100	
India	1 186 200 000	1 040	1.6	67	63	61	1.3
Indonesia	234 300 000	2 180	1.3	70	67	91	1.9
Iran	72 200 000	5 250	1.0	73	70	77	6.5
Iraq	29 500 000			62	59	74	
Ireland	4 400 000	64 660	0.9	81	76	100	10.2
Israel	7 000 000	26 535	1.5	83	78	97	9.2
Italy	58 900 000	40 450	0.1	84	77	99	7.7
Ivory Coast	19 600 000	1 250	2.0	47	46	50	0.5
Jamaica	2 700 000	4 990	1.0	73	69	80	3.8
Japan	127 900 000	37 940	-0.1	86	79	100	9.6
Jordan	6 100 000	3 270	2.4	74	71	90	3.8
Kazakhstan	15 500 000	9 075	0.8	70	59	100	11.9
Kenya	38 600 000	890	2.7	49	51	74	0.3
Kiribati	100 000	710					

STATE	NUMBER OF INHABITANTS	GNP/CAPITA US DOLLARS	POPULATION* INCREASE/Y, %	LONGEVITY FEMALE	LONGEVITY MALE	LITERACY %	CO_2/CAPITA TONS/YEAR**
Korea, Dem. P. Republic (North)	23 900 000		0.5	67	61		3.5
Korea, Republic (South)	48 400 000	19 640	0.4	82	74		9.4
Kuwait	2 900 000	46 400	1.7	80	76	93	36.9
Kyrgyzstan	5 400 000	950	1.4	72	64	99	1.1
Kosovo	2 130 000	1 800				94	
Laos	6 000 000	830	2.7	58	55	69	0.3
Latvia	2 300 000	14 930	-0.6	78	67	100	2.8
Lebanon	4 100 000	7 375	1.1	75	71		4.2
Lesotho	2 000 000	675	0.9	34	34	82	
Liberia	3 900 000	235	3.1	43	42		0.1
Libya	6 300 000	17 470	1.9	77	72		9.5
Liechtenstein, Principality of	34 000					100	
Lithuania	3 400 000	14 460	-0.4	79	68	100	4.1
Luxembourg	472 000	118 045				100	
Madagascar	20 200 000	480	2.8	57	55	71	0.2
Macedonia	2 000 000	4 685	0.2	77	72	96	5.1
Malawi	14 300 000	230	2.8	40	41	65	0.1
Malaysia	27 000 000	7 870	1.7	76	72	89	9.3
Maldives	311 000	3 760					
Mali	12 700 000	655	3.4	50	48	20	0.05
Malta	408 000	20 745					
Marshall Islands	81 000						
Mauritania	3 200 000	1 195	3.2	56	53	52	0.6
Mauritius	1 300 000	6 390	0.7	76	70	85	2.7
Mexico	107 800 000	10 750	1.3	79	74	91	4.1
Micronesia Federation	540 000	2 300					
Moldova, Republic of	3 800 000	1 830		73	66	99	2.1
Monaco	33 000	30 000				100	

STATE	NUMBER OF INHABITANTS	GNP/CAPITA US DOLLARS	POPULATION* INCREASE/Y, %	LONGEVITY FEMALE	LONGEVITY MALE	LITERACY %	CO_2/CAPITA TONS/YEAR**
Mongolia	2 700 000	1 880	1.2	68	64	98	3.4
Montenegro	600 000	4 150		74	74	97	
Morocco	31 600 000	2 900	1.7	73	69	53	1.6
Mozambique	21 800 000	470	1.9	42	42		0.1
Myanmar/Burma	49 200 000	230	0.9	65	59	90	0.2
Namibia	2 100 000	3 800	0.9	45	46	85	1.3
Nauru	13 800						
Netherlands	16 500 000	54 445	0.4	82	76	100	7.7
Nepal	28 800 000	460	2.1	64	63	49	0.1
New Zealand	4 300 000	31 700	0.7	82	78	100	7.2
Nicaragua	5 700 000	1 060	2.3	73	69	77	0.2
Niger	14 700 000	385	3.9	45	45	29	0.1
Nigeria	151 500 000	1 490	2.4	44	44		0.8
Norway	4 700 000	102 525	0.3	83	78	100	11.4
Oman	2 700 000	21 700	2.2	77	74	81	12.5
Pakistan	167 000 000	1 000	1.9	65	64	50	0.9
Palau	20 800	8 400	0.7	75	68		
Panama	3 400 000	6 880	1.7	78	73	92	1.8
Papua New Guinea	6 500 000	1 030	1.9	58	56	57	0.7
Paraguay	6 200 000	2 660	2.3	74	70		0.7
Peru	28 200 000	4 600	1.6	74	69	88	1.4
Philippines	89 700 000	1 910	1.9	74	69	93	0.9
Poland	38 000 000	14 890	-0.1	79	71		7.9
Portugal	10 700 000	24 030	0.1	81	75		5.9
Qatar	856 000	106 460					
Romania	21 300 000	9 950	-0.2	76	69	97	4.1
Russian Federation	141 800 000	12 580	-0.4	72	59	100	10.5
Rwanda	10 000 000	420	2.4	46	43	66	0.1
Saint Kitts & Nevis	50 000	10 560	0.9				

STATE	NUMBER OF INHABITANTS	GNP/CAPITA US DOLLARS	POPULATION* INCREASE/Y, %	LONGEVITY FEMALE	LONGEVITY MALE	LITERACY %	CO_2/CAPITA TONS/YEAR**
Saint Lucia	160 000	6 064					
Saint Vincent and the Grenadines	110 000	5 580					
Samoa	189 000	2 800	2.2	75	69	100	
San Marino	30 000	40 000				99	
São Tomé & Principe	160 000	1 000					
Saudi Arabia	25 300 000	21 220	2.3	75	71	78	16.5
Senegal	12 700 000	1 110	2.7	58	56	40	0.4
Serbia	9 900 000	7 060	-0.3	73	73	97	6.5
Seychelles	85 000	9 440					
Sierra Leone	6 000 000	335	2.4	43	40	36	0.2
Singapore	4 500 000	41 290	0.6	81	78	93	13.2
Slovakia	5 400 000	18 585	0.0	79	71	100	6.8
Slovenia	2 000 000	28 330	0.0	81	74		7.4
Solomon Islands	507 000	880					
Somalia	9 000 000		2.7	50	47		0.1
South Africa	48 800 000	6 170	0.3	44	44	83	8.7
Spain	44 600 000	36 970	0.2	84	77	100	7.9
Sri Lanka	19 400 000	2 100	1.2	78	72	91	0.6
Sudan	39 400 000	1 630	2.2	58	56	62	0.3
Suriname	461 000	5 600					
Swaziland	1 100 000	2 900	1.2	29	31	80	0.8
Sweden	9 200 000	55 620	0.0	83	79	100	5.4
Switzerland	7 500 000	67 380	0.2	84	78	100	5.5
Syrian Arab Republic	20 400 000	2 240	2.5	76	72	80	3.6
Tajikistan	6 800 000	740	2.2	67	62	100	0.8
Taiwan	23 000 000	18 310					
Tanzania United Republic	41 500 000	520	2.3	47	46	70	0.1
Thailand	64 300 000	4 100	0.8	75	68	93	4.3

STATE	NUMBER OF INHABITANTS	GNP/CAPITA US DOLLARS	POPULATION* INCREASE/Y, %	LONGEVITY FEMALE	LONGEVITY MALE	LITERACY %	CO_2/CAPITA TONS/YEAR**
Timor-Leste	1 200 000	460		59	56		0.2
Togo	6 800 000	455	2.8	57	54	54	0.2
Tonga	102 100	2 510	1.7	73	68	100	
Trinidad & Tobago	1 300 000	18 865	0.6	73	68	98	24.7
Tunisia	10 400 000	4 030	1.1	76	72	74	2.2
Turkey	75 800 000	11 465	1.3	72	67	88	3.4
Turkmenistan	5 000 000	4 180	1.4	67	59	99	8.6
Tuvalu	12 200						
Uganda	31 900 000	470	3.7	52	51	68	0.1
Ukraine	45 900 000	4 320	-0.7	73	61	100	6.9
United Arab Emirates	4 500 000	56 670	1.5	82	77		30.1
United Kingdom	61 000 000	45 680	0.2	81	77	100	9.1
Uruguay	3 400 000	8 860	0.6	80	73		1.7
USA	308 800 000	47 025	0.6	81	75	98	19.5
Uzbekistan	27 800 000	980	1.4	70	64		4.3
Vanuatu	232 000	2 385					
Vatican City, State of the	1 000					100	
Venezuela	28 100 000	11 830	1.7	77	71	93	5.6
Vietnam	88 500 000	1 050	1.3	74	70	91	1.2
Yemen	23 100 000	1 200	3.2	64	61		1.0
Zambia	12 200 000	1 220	2.1	38	39	68	0.2
Zimbabwe	13 500 000	1 380	1.1	36	38		0.9

Source: Swedish Institute of International Affairs, database, 2009;
**) Refers to natural increase of population, excluding immigration and emigration.*
***) Data in last column refers to carbon dioxide emissions per capita (CO2/capita) for year 2005.*
All tons are metric.